The Earth…

…but not
As We Know It

An Exploration

Andrew Johnson

Book details:

Title: The Earth… but not As We Know It An Exploration

by Andrew Johnson

Self-Published – October/November, 2019

ISBN-13: 9781702017572

Front Cover Design by Andrew Johnson, based on a "Canva/Pixabay" image by "Comfreak"[1]

Back Cover Image: Pixabay[2]

Reference List:

Only 1 Wikipedia link included! Yikes!

(Wikipedia censors important information in real-time - use it at your peril!)

Use of Images

No copyright infringement or misuse of images is intended in this book, which is designed to educate and enlighten, using criticism and analysis. A version of this book will be given away as a free download. However, if you find an image in this book you are concerned about, please contact me and if you feel uncomfortable giving me permission, I will certainly remove it, if that is your wish. Thank you for your understanding!

Acknowledgements

Thanks to everyone that has helped in developing this research and to those who make the information, herein, freely available.

Special Acknowledgements

Fredrik Nygaard - for his very useful comments and corrections in several parts of this book.

- Dr James Maxlow
- Stephen Hurrell
- Peter Woodhead
- Fredrik Nygaard

Important Researchers Quoted

- Dr James Maxlow
- Peter Woodhead
- Stephen Hurrell
- Jan Lamprecht
- Neal Adams
- Fredrik Nygaard
- Dr James Maxlow
- Wallace Thornhill

Author Biography

Andrew Johnson grew up in Yorkshire and graduated from Lancaster University in 1986 with a degree in Computer Science and Physics. He worked in Software Engineering and Software Development, for about 20 years. He has also worked full and part time in lecturing and tutoring and he now works for the Open University (part time) tutoring students, whilst occasionally working on various small software development projects. He became interested in "alternative knowledge" in 2003, soon after discovering Dr Steven Greer's Disclosure Project. Andrew has given presentations and written and posted many articles on various websites about 9/11, Weather Anomalies, Solar System Anomalies, and Antigravity research, and he has occasionally challenged some of the authorities to address some of the data he and others have collected.

Andrew is married and has two children. You can email Andrew Johnson on ad.johnson@ntlworld.com.

His website is www.checktheevidence.com.

Other Books by Andrew Johnson

These are available as PDF downloads, as well as other formats:

Climate Change and Global Warming… Exposed: Hidden Evidence, Disguised Plans (ISBN: 978-1976209840)

This book collects together, for the first time anywhere, a range of diverse data which proves that the whole issue of "climate change" is more complicated and challenging than almost all researchers are willing to consider, examine, or entertain. For example, this book contains astronomical data which most climatologists will not discuss in full. Similarly, the book contains climate and weather data that astronomers will not discuss. The book contains some data that neither astronomers nor climatologists will discuss. It contains some data that no scientists will appropriately discuss. It will show the reader why the climate change/global warming scam was invented, and it will illustrate how the scam has been implemented.

Secrets in the Solar System: Gatekeepers on Earth (ISBN: 978-1981117550)

Since 1957, robotic space probes have visited all the planets in the Solar System. Is it the case that they have only found mostly uninteresting collections of gas, rocks, ice and dust? Has any evidence of past or present life in the Solar System ever been discovered?

This book will take you "on a journey" to the Sun and the Moon, Mars, Phobos, Saturn and some of its moons. It will show you some of the numerous anomalies that have been found. Could it be the case that taxpayer-funded space agencies have ignored or even lied about these anomalies, and their significance?

After more than 10 years of ongoing research, collected together here, for the first time anywhere, are over 350 fully referenced pictures and data from over 50 years of space missions. Anomalous images are presented with some detailed explanations, commentary and analysis, completed by various researchers. The book asks what would happen if NASA or ESA scientists had discovered compelling evidence of past or present extra-terrestrial life in the Solar System? Would they "tell us the truth, the whole truth and nothing but the truth" about such a discovery? Or, would the "scientific technological elite" mentioned in Eisenhower's final address to the USA in 1961, become the "gatekeepers" of "Secrets in the Solar System?"

Finding the Secret Space Programme: Removing Truth's Protective Layers (ISBN: 978-1981283705)

The "Space Age" began in 1957 and since then, we are told, missions have studied all the planets in the Solar System. Most people believe that one of the greatest technological achievements happened in 1969, when Neil Armstrong allegedly walked on the surface of the moon. The last Apollo mission ended in 1972 and later manned space missions never travelled beyond low Earth orbit, despite enormous advances in many technologies. This book considers the works of Dr Paul La Violette and Defence Journalist Nick Cook regarding antigravity research in the aerospace industry and elsewhere. It analyses the statements of certain whistle-blowers, who claim to have worked on covert propulsion and energy technology programmes. The Apollo programme and its background are also studied in some depth, along with claims and statements made by some of the Apollo astronauts.

Is it true that exotic propulsion technologies have been successfully developed and deployed, as part of a secret space programme - and comprehensively hidden from public view, using sophisticated methods?

After more than 14 years of ongoing research, collected together here, probably for the first time anywhere, are referenced images, quotes analysis and commentary. A study of this evidence should help us to remove what Neil Armstrong referred to as "Truth's Protective Layers."

9/11 Finding the Truth (ISBN: 978-1548827618)

What really happened on 9/11? What can the evidence tell us? Who is covering up the evidence, and why are they covering it up? This book attempts to give some answers to these questions and has been written by someone who has become deeply involved in research into what happened on 9/11. A study of the available evidence will challenge you and much of what you assumed to be true.

9/11 Holding the Truth (ISBN: 978-1979875981)

The truth about what happened to the World Trade Centre on 11 September 2001 was discovered by Dr Judy Wood, through careful research between 2001 and 2008. The author of this book, Andrew Johnson, had a "good view" of how later parts of Dr Wood's research "came together." Not only that, he was also involved in activities, correspondence and research which illustrated that this truth was being deliberately covered up. This book is a companion and follow-up volume to "9/11 Finding the Truth" - and documents ongoing (and successful) efforts to keep the truth out of the reach of most of the population. Evidence in this book, gathered over a period of 12 years, shows that the cover up is "micro-managed," internationally and even globally. The book names people who are involved in the cover up. It illustrates how they often stick to "talking points" and seem to have certain patterns of behaviour. It attempts to illustrate how difficult it is to prevent the truth from being marginalised, attacked and "muddled up."

Acknowledged: A Perspective on the Matters of UFOs, Aliens and Crop Circles (ISBN: 9781726690911)

Have aliens really visited us? What do they look like? Where do they come from and why are they here? Have they given us messages in Crop Circles? What historical evidence is there that aliens have been involved in the evolution or creation of humans? Is there a UFO/Alien cover up? If so, how is it kept in place? This book embodies an attempt to give some answers to the questions above, based on a compilation and distillation of evidence collected during about 15 years of ongoing research. Although this book is not written for people who are new to these topics, "newbies" and seasoned researchers alike should find something useful herein. Detailed information and over 900 references aim to give the reader a perspective on alien intervention, alien abduction, alien contact, landings and crashes. The involvement of military and intelligence interests and their influence on the so-called "Disclosure" movement is also considered. Also included is a critical look at certain prominent UFO researchers, who seem to be helping to keep certain truths obscured or marginalised. The second part of the book studies various facets of the Crop Circle Phenomenon in a similar way to part one.

Although most or all of the information in this book has been written about elsewhere, this work attempts a synthesis of reasoned analysis which, it is hoped, will enlighten the reader and give them a new understanding about why any official type of "Disclosure" is unlikely to happen, as powerful interests need to keep their crimes covered up.

Table of Contents

Preface and Introduction

This book is a little different from the others I have written. Some readers of my earlier works may think that what is in this book isn't nearly as serious or as important as the content of my other books. If they think that, they are almost certainly correct. However important the contents of my earlier books, their effect will probably still be fairly limited. Those books were written because of my initial curiosity about - and inquiry into - narratives related to other issues which I wanted to find the truth about. Again, it is my curiosity which led me to write this book - where, again, I am questioning a commonly stated narrative and/or conclusion. In this book, however, the matter being analysed is (primarily) geological and even geographical in nature!

Why This Book Exists

As I began to investigate other areas of "alternative knowledge," it wasn't very long before I came across videos by a chap called Neal Adams. I also heard him discuss these same videos on an edition of Coast to Coast. The videos illustrated - very clearly - an intriguing observation: the Earth had been slowly expanding, over millions of years. The video presentation was very compelling and the basis of it seemed, to me, irrefutable. However, as geology and Earth sciences aren't something I had previously studied in depth, I didn't feel I had anything to add to the presentations that Neal Adams had already made available. Sometime after this, I came across the research of an Australian geologist, Dr James Maxlow. I could see that Maxlow's comprehensive work essentially grounded Adams' illustrations and assertions in some scientific evidence, which convinced me that the Earth has, indeed, expanded since its original formation and it is still expanding today.

In 2012, I was asked to do a presentation for a group and this gave me "an excuse" to collate and format some of the "Earth expansion" topics and evidence I'd come across up to that time. I decided that if I was going to engage an audience for two hours, I would need an additional topic for the presentation, so I chose a related topic that some people ridicule even more frequently than the idea of an expanding Earth - and that is the idea of a hollow Earth! Why would I even want to cover such "nonsense?"

This is a fair question and the answer to it, for me at that time, was quite straightforward. I'd seen an intriguing video about the stories of travelling to the "inner Earth" - inside the "hollow Earth" and I decided that this was worthy of some more detailed research. In the end, this proved to be much more worthwhile than I initially expected - later "events" led to the production of this book.

Having recorded the audio of my "Earth" presentation, which I gave to a group of about 10 people in Leicester, I then simply synchronised the recording to the PowerPoint slides and edited in some of the video clips I had used. I uploaded this "Hollow Earth" video to YouTube," where it became one of the most popular on my channel - garnering 80,000 views in a few months - easily twenty times more than most of my other channel's videos! I was surprised that these topics had proved so popular. Perhaps the popularity was due to the title I chose for the video - "The Earth But Not As We Know It." I am really not sure where all the views came from - as I didn't promote the video and never received much correspondence from anyone about it. The video, however, was blocked in 2017 or 2018, due to a copyright claim on one of the video clips I had used from the Michael Palin "Pole to Pole" series (discussed later). The original video, however, is still available on "Daily Motion," should you wish to view it[3] (though some of the information in it is out of date now)!

Some time after I posted this video, I was contacted by a property developer, Peter Woodhead who lives in Lancashire, UK. He had obviously watched the video! He said that he had a new explanation for the way in which the Earth had expanded. In essence, his explanation/theory neatly connected the two halves of my 2012 presentation! I was intrigued!

Further to this, we realised that there were other implications of this explanation which seemed to tie in with the so-called "Electric Universe" (EU) model - as proposed, in the modern era, primarily by Australian Physicist Wallace Thornhill. Following our posting of articles relating to Peter Woodhead's Earth expansion explanation, we were contacted by Fredrik Nygaard who then spent a considerable time developing further aspects of the explanation which seemed to strengthen the connection to the afore-mentioned EU model. The end result was a series of postings and articles by Peter Woodhead and Fredrik Nygaard which primarily appeared at my website - http://www.checktheevidence.com/.

This book, then, represents a compilation and distillation of this information - with updates, corrections and augmentations, where applicable. Some of the conclusions are speculative, but others are quite firm, so please read on!

Maths and Science

Though this book is not a "hard science" book, it does contain some discussion of scientific concepts and data. Most conventional scientists will dismiss most of what is written here, for various reasons, but those who read through, and remain open-minded, might find errors that are worth correcting. If this is the case, please do contact me on ad.johnson@ntlworld.com. Similarly, when discussing the evidence relating to the Earth's expansion, I will be showing a fair few calculations. These are mainly volume and force related calculations. These have been checked but, again, if any reader finds errors, please forward corrections, as stated above. I appreciate some readers may find these calculations a bit hard to follow, but they are provided for those who wish to understand how certain conclusions were reached.

Part 1

A Hollow Earth?

1. Our Spherical Home Planet

Introduction

Observations of various kinds - both from the ground and from above ground - show our home planet to be a sphere. Measurements made over 2000 years ago by Eratosthenes of Cyrene, (276 BC - 194 BC), the Greek scientific writer, Astronomer and Poet[4], showed that the radius of the Earth was in the region of 6000km. (Sufficiently interested people could easily repeat the measurements made by Eratosthenes, if they wished to verify his results!) Today, the actual figure is known to be close to 6370km.

But, curious people (like me) may then ask questions such as "what lies beneath the surface of the Earth?" and "what would we find if we went down through a volcanic vent, further and further towards the centre of our planet?" Related questions then might be "how did our Earth form?" and "When did it form?"

Of course, "mainstream" science is considered by many to have found all the important answers to these questions, which we will cover in the chapters of this book. We will also cover a few aspects of Earth's geological and biological history which mainstream science does not seem able to explain.

Development of the Earth - The Accepted Theory

Formation

Much of what is discussed below is derived from an article by Raymond Jeanloz and Jonathan I. Lunine available on Britannica.com[5].

It is thought that the Earth was initially formed out of a cloud of gas and dust which made up the early solar system. The article above explains:

> *Under its own gravitational attraction, the cloud collapsed into a rotating disk of matter, called the solar nebula.*

This brings us to the first assumption - that gravity and rotation were the primary forces or processes which originally created the Earth from pre-existing elements. However, it seems obvious that the angular momentum for this rotation must have come from somewhere, so the article then states:

> *The collapse could have been initiated by a shock wave emanating from a nearby supernova, a violently exploding star, or by random density fluctuations in the cloud itself.*

The article then goes on to discuss the formation of the Sun and the Planets, at more or less the same time - approximately 4.5 billion years ago. The dating of the formation of the Earth has been estimated based on the decay of radioactive isotopes, currently found in many rocks. This same area of study has led some scientists to conclude that a proportion of the Earth's mass is made up of "solid material from outside the solar system." Scientists and researchers have studied the composition of the Sun (mainly using spectroscopy) as well as the elements found in many meteorites - specifically carbonaceous chondrites[6]. They have considered how the composition of these "fallen stars" can be compared with the composition of rocks found here on Earth. The article describes:

> *This is the basis for the chondritic model, which holds that Earth (and presumably the other terrestrial planets) was essentially built up from bodies made of such meteoritic material. This idea is corroborated by isotopic studies of rocks derived from interior regions of Earth considered to be little changed throughout the planet's history. Thus, it appears that the composition of Earth is roughly what would be expected given the observed elemental abundances in the Sun and accounting for the loss of the more volatile elements.*

The formation of the Earth must have happened once the gas and plasma had somehow condensed - after cooling to form solid particles or grains. The article mentions accretion and gravity a few times when trying to explain the formation of the Earth and other bodies in the solar system. It then mentions the temperatures that may have been apparent in the early formation of the Earth (which will be of particular interest to us later).

> *Thus, a wide range of minerals was included in the grains, the larger fragments, and even the planetesimals that were accumulated by the growing planet. Apparently, such an aggregation of dense metallic fragments and less dense rocky fragments is not very stable. Calculations based on the measured strengths of rocks indicate that the metallic fragments **probably sank downward** as Earth grew. Although the planet was relatively cold at this stage - less than 500 K (440 °F; 230 °C) - the rock was weak. This is an important point because it **leads to the conclusion that Earth's metallic core began to form during accretion** of the planet and probably before the planet had grown to one-fifth of its present volume.*

So, it is assumed that the Earth developed a metallic core - because the metal elements which were present in the primordial mixture were the heaviest and therefore "sank" to the centre. Whilst intuitively, this seems like a good explanation, we will question this later.

Development

It is then suggested in the article above, and in most other sources attempting to explain how the Earth formed, that planetesimals (large meteors, asteroids and other giant rocks) then played a great role in the development of the Earth:

> *During its accretion, Earth is thought to have been shock-heated by the impacts of meteorite-size bodies and larger planetesimals. For a meteorite collision, the heating is concentrated near the surface where the impact occurs, which allows the heat to radiate back into space. A planetesimal, however, can penetrate sufficiently deeply on impact to produce heating well beneath the surface. In addition, the debris formed on impact can blanket the planetary surface, which helps to retain heat inside the planet. Some*

scientists have suggested that, in this way, Earth may have become hot enough to begin melting after growing to less than 15 percent of its final volume.

We can pause and note that this is all theoretical - with no specific evidence to prove that the suggested processes and events really did effect the Earth's formation in the way suggested.

When considering the formation of the Earth, it is, of course, quite logical to consider the formation of the moon too - and also observe the moon's size, relative to the Earth. That is to say that, compared to all other bodies in the solar system, the Earth (relatively speaking) has the largest moon i.e. Jupiter has a mean diameter of 137,346km and its largest moon, Ganymede has a diameter of 5262km - so the ratio of these 2 bodies is 0.038. The Earth has a diameter of about 12742 km and the moon has a diameter of about 3474 km - so the ratio of these 2 bodies is 0.27.

This, I would argue, is difficult to explain, but the current thinking relating to this issue is covered in the "Britannica" article mentioned above:

Among the planetesimals striking the forming Earth, at least one is considered to have been comparable in size to Mars. Although the details are not well understood, there is good evidence that the impact of such a large planetesimal created the Moon. ***Among the more persuasive indications is that the relative abundances of many trace elements in rocks from the Moon are close to the values obtained for Earth's mantle.***

(At this point, I must suggest that if this article is referring to rocks brought back in the Apollo missions, we cannot be certain of the data from those - because no Apollo astronauts landed on the moon. It is possible, however, that rocks that were meteorites[7] could have been used to determine these compositions, or that the rocks were brought back to the Earth by some other undisclosed method. I know any scientifically minded people will be shocked at my statement - but, it is true. Please study my "Finding the Secret Space Programme" book[8] to learn more.)

The "Mars-sized-object collision with the Earth" theory has been around for at least 30 years, but I have never been particularly comfortable with it - due to the kinetic energies involved, which would seemingly cause a total disintegration of the bodies concerned. That said, I have not studied the details of this theory, nor the calculations on which it is based. However, it seems to me that a more logical suggestion would be that the Earth's mantle and the moon formed at the same time, from some other process - not a collision or "excavation," as the article goes on to suggest. Despite this enormous excavation, the article suggests the core continued to grow:

Simultaneously, Earth's core was accumulating and may have been completely formed during the planet's growth period. In addition to the possible accretional heating caused by planetesimal impacts, the sinking of metal to form the core released enough gravitational energy to heat the

Hence, it is assumed that the core of the Earth became much hotter after its initial formation. This is an idea we will revisit later in some detail, along with the issue of what is called "plate tectonics," which is the study of how the landmasses/continents formed and have moved around on the Earth's surface.

Simplified Representation of Structure of Earth

The diagram below shows the proposed structure of the Earth - and this picture has remained the same for at least 50 years (I remember seeing similar images in a children's encyclopaedia I had):

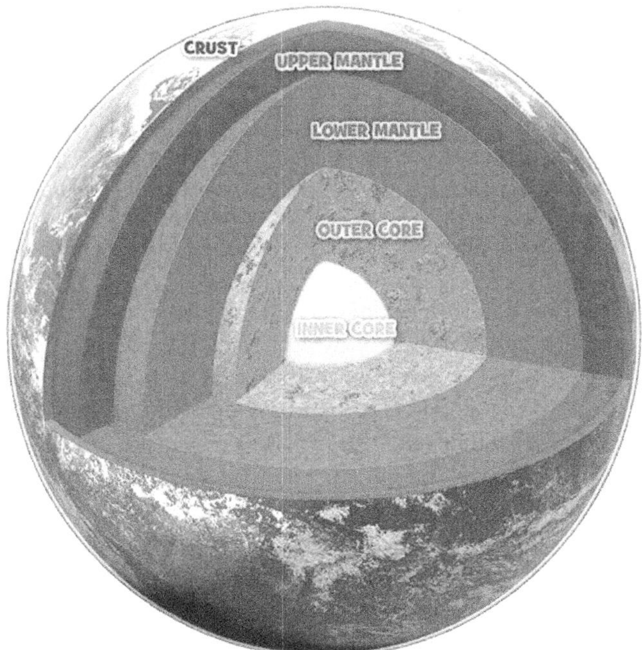

The image above came from a National Geographic website[9], which also describes the structure of the Earth in a little more detail. It explains that the inner core is thought to be made of iron and nickel and have a temperature of between 5000°C and 6000°C, while the outer core is made of iron, nickel, sulphur and oxygen. The lower and upper mantles are both made of iron, oxygen, silicon, magnesium and aluminium and are thought to be at temperatures of 3000°C and 1400°C to 3000°C respectively. Hence, it is shown/suggested that the temperature of the material layers increases as one moves towards the centre - where the density of material and pressure is also increased.

The outer layer - known as the crust - is between 8km (oceanic crust - of mostly basalt) and 70 km thick (continental crust - of mostly granite).

Also described, in the article referenced above, is how the structure of the Earth has been determined from seismic studies, when waves produced during earthquakes or volcanic eruptions are measured by various sensors, positioned around the globe. The measurement of the way these seismic signals travel and change as they move through the various layers of material can be used to determine something about the composition of each layer. Again, we will consider the Earth's structure and seismic signals in more detail, later.

The Earth's Magnetic Field

Another important characteristic of our home planet - unlike some other planets in the solar system - is that it has a magnetic field. The reason for the presence of our magnetic field is not really all that clear, though the Britannica article[10] has this to say about why it is extant:

> *Helical fluid motions in Earth's electrically conducting liquid outer core have an electromagnetic dynamo effect, giving rise to the geomagnetic field. The planet's sizable, hot core, along with its rapid spin, probably accounts for the exceptional strength of the magnetic field of Earth compared with those of the other terrestrial planets. Venus, for example, which has a metallic core that may be similar to Earth's in size, rotates very slowly and has no detected intrinsic magnetic field. Mercury and Mars have only small intrinsic magnetic fields.*

Hence, it is suggested that the flowing metal in the core also has electrical currents flowing through it and the flowing electrical currents, in turn, create a magnetic field. This assumes that the flow is coherent enough to create a relatively uniform magnetic field. That is to say that the currents in the liquid core don't "cancel each other out." It is difficult to explain why the Earth's magnetic field is now weakening - as the core should have no reason to change and we know the rotation speed of the Earth (which dictates the length of a day!) has not significantly changed. The article, above, assumes that planetary rotation and core structure are what create the magnetic field. This is why it just suggests that Mars, Venus and Mercury don't have such a strong magnetic field, either because their planetary cores are different to the Earth's, or their rotation speeds are different to the Earth's.

Formation of the Continents

Another matter for consideration is how the continents we know today came into being. In the early 1900s Alfred Wegener proposed the idea of Continental Drift[11]. He suggested that continents moved across the face of the Earth. This idea, initially ridiculed and dismissed as pseudoscience - was ultimately developed into what is known as plate tectonics[12] today. Before this, however, when more accurate continental maps were developed, near the end of the 18th century, it was noticed that the lands bordering the Atlantic Ocean looked like they had once been joined into a larger continent.

One of the first people to suggest that this was actually the case was Alexander von Humboldt - a German naturalist.

Later, it became evident that the same species of plant and animal life were found in the fossil record - on continents now thousands of miles apart. A similar story is told by some of the geology that has been studied.

It is really only since the 1950s that this view has become accepted. Since then, a timeline for the break-up of a so-called "super-continent" has been proposed, which has resulted in the continents we have today. This is shown in the sequence of diagrams below from "Geology.com"

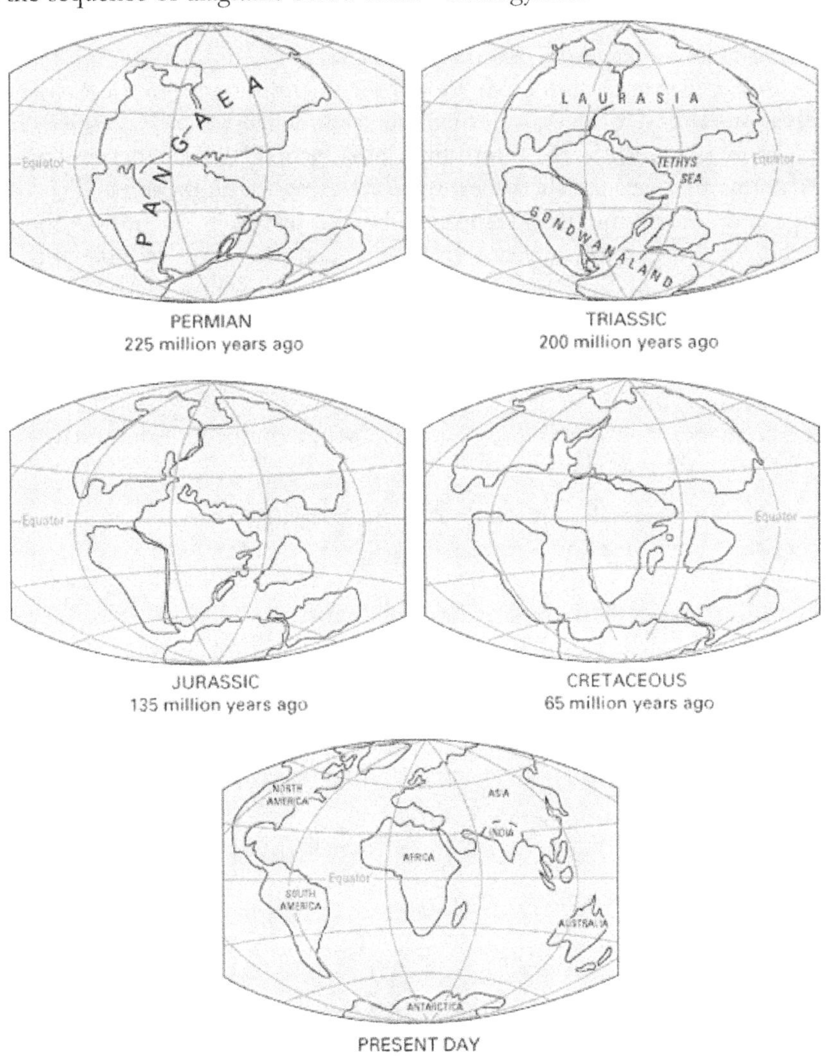

PERMIAN
225 million years ago

TRIASSIC
200 million years ago

JURASSIC
135 million years ago

CRETACEOUS
65 million years ago

PRESENT DAY

From the Geology.com website, we can read:[13]

> *Plate tectonics is the study of the lithosphere, the outer portion of the Earth consisting of the crust and part of the upper mantle. The lithosphere is divided into about a dozen large plates which move and interact with one another to create earthquakes, mountain ranges, volcanic activity, ocean trenches and many other features. Continents and ocean basins are moved and changed in shape as a result of these plate movements.*
>
> *The sequence of maps on this page shows how a large supercontinent known as Pangaea was fragmented into several pieces, each being part of a mobile plate of the lithosphere. These pieces were to become Earth's current continents. The time sequence shown through the maps traces the paths of the continents to their current positions.*

The article also notes the differences in the spelling of "Pangea "stating that "Pangaea" is the preferred spelling. Notice that in all the diagrams above, the continents, whatever their size, are surrounded by ocean. Also note that *only trans-Atlantic* joining/splitting of continental masses is considered. The split across *the Pacific* is not considered. We will see the importance of these observations later. However, we can immediately suggest that the east to west split is more obviously in the forefront of most people's minds, because of the way most world maps are shown - with North America on the left and Europe and Asia on the right. Hence, the brain is less likely to consider the proximity of, for example, the eastern Russian continent to Alaska (i.e. the west to east split). So the Pangea models are "driven" by how maps have been drawn for decades or centuries, rather than being driven by the way, for example, the Earth can be observed from space.

We will see later that these proposed maps actually contradict available geological evidence, in some cases, and this "supercontinent" cannot have fragmented in the way shown.

Also notice the timescales involved - millions of years. We may later use the abbreviation "mya" for "million years ago."

Questioning Aspects of the Formation Theory.

While we can easily consider that the explanation for the formation of the Earth is "good enough," there are a number of questions which are raised when we consider more details. These are:

- How did Pangea separate and become the continents we know today?
- How did the deep oceans form?
- Has the Earth always been the same size, during the period in which complex life has existed here?

Losing All Credibility?

Now that we have covered the accepted theories about how the Earth formed and what the internal structure might be, let us consider some ideas which most scientists find ludicrous (sometimes for good reason). The reason for doing this is firstly to try and dispel some myths, but then also to illustrate that studying "crazy" ideas carefully may help us to find new explanations that seemingly fit available evidence better than more conventional and accepted explanations/theories. Please stick with me!

2. Stories of A Hollow Earth

Pure Fantasy!

It was a long time ago that I first came across the idea of the Earth being hollow - probably as a child, when I saw the film "At the Earth's Core" starring Doug McClure and Peter Cushing[14]. Indeed, most people only consider the idea for as long as the film/movie lasts, or for the length of time it takes them to read novels such as "Journey to the Centre of the Earth[15]." And, of course, the idea is only acceptable when being considered for its entertainment value!

The idea of a Hollow Earth was written about by Edgar Rice Burroughs (creator of the Tarzan stories) in a series of 7 novels, commencing with the 1914 novel "At the Earth's Core."[16] In the first story, the heroes drill down into the Earth a distance of 500 miles and thereby discover a prehistoric world below the surface. The inner Earth is constantly illuminated by an unsetting inner sun. However, there is also an "inner moon" which creates a "Land of the Dreadful Shadow" on the inner surface. Burroughs includes the idea of openings at the poles, through which Zeppelin dirigibles travel to the fantastic interior land.

I was reminded of the Hollow Earth idea several times in the course of my "web travels." Once such reminder was in the form of a short, enjoyable YouTube video by Nies Lighting[17]. This video triggered my attempts to investigate the idea further. I hope that, when the reader studies the later sections of this book, they will realise why I decided to expend a fair amount of effort in attempting to unravel some of the accounts related to the "Hollow Earth."

It may surprise some readers, then, to learn that at least one very well-known scientist (or rather "Natural Philosopher") considered the idea of a hollow Earth seriously…

Early History of the "Hollow Earth"

Early Myths and Legends

An interesting page on the "Crystal Links" website [18]by an anonymous author contains some useful information about how the idea of a Hollow Earth has been promulgated.

We can start by considering the idea of "subterranean realms" which almost certainly would have been first considered when men descended into cave systems around the world. We can then move on to consider more complex mythologies and beliefs that were found in, for example, the Greek Classics - where Hades is God of the Underworld[19]. In Norse mythology, there is Svartalfheim [20]("Homeland of the Black Elves") and "Nidallevir" (the "Dark Fields") which is said to be where the Dwarves live - beneath the ground, or perhaps in the "northern lands".

One might even consider the "Christian Hell," as being a reference to a "hot place," beneath our feet - where people can be sent to suffer. Some would align this concept with the Jewish "Sheol" [21] which apparently means "Abode of the Dead."

In Kabbalistic literature, such as the Zohar[22], reference is made to "a land below" called Arka:

> *…Arka is one of the seven lands below, where the sons of Cain's sons reside. Indeed, after being banished from the face of the Earth, they*

Some have associated the mythical "Shambhala," spoken of in Buddhist teachings, with the Inner or Hollow Earth[23]. It was said to be an underground or isolated kingdom in the Himalayas.[24]

Post Renaissance and Western Culture References

In 1692, Edmond Halley (pronounced Hal-ee, or Hall-ee - not Hay-lee) published an article that proposed that the Earth might be made up of a series of shells. Halley was an astronomer, geophysicist, mathematician and meteorologist - he worked out the orbit of a particular comet - which he said had a periodicity of 76 years. The comet did return and was then named after him.

An excellent article, written in 1992, by Dr Nick Kollerstrom[25] gives many more details about Halley's theories, which were seemingly triggered by his observations related to both the Earth's magnetic field (measured with a compass) and the relative sizes of the Earth and the moon. The latter considerations related to his friend Isaac Newton's "Principia" work on gravity, which Halley was instrumental in getting published. Kollerstrom writes:

Halley's Earth was composed of an outer shell 500 miles thick, with an air gap of the same distance between it and the inner sphere. To the objection that the latter might collide with the outer shell, and thereby damage it, he explained that it was held at the centre by the force of gravity. Halley was confident his readers would perceive the necessity of this: "should these globes be adjusted once to the same common centre, the Gravity of the parts of the Concave would press towards the centre of the inner ball ... it follows that the Nucleus being once fixt in the common centre, must always here remain. Halley pointed out that "the Ring environing the Globe of Saturn", which remained coaxial to the planet, was held there by gravity. (No-one then knew that Saturn's rings were rotating. The Principia had not discussed the matter.) By analogy, could not gravity also hold a globe concentric inside the hollow Earth?

In his 1692 paper, called "An account of the cause of the change of the variation of the magnetical needle with an hypothesis of the structure of the internal parts of the Earth"[25] Halley's suppositions go further, where he writes:

What curiosity in the structure, what accuracy in the mixture and composition of the parts ought we not to expect in the fabric of this Globe, made to be the lasting habitation of so many various species of animals, in each of which there want not many instances that manifest the boundless power and goodness of their Divine Author…

Hence Halley was postulating these inner shells or areas might be inhabited! Also in the same paper, he supposed that the surfaces of the shells would be luminous.

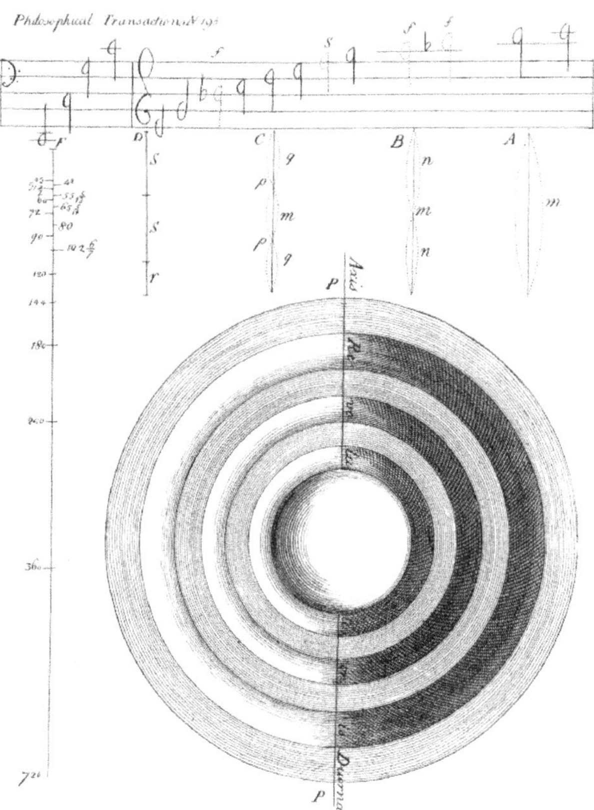

Illustration from Halley's 1692 paper [26] which considered the idea of a (partly) Hollow Earth

Some web postings claim[27] that the prodigious Russian Mathematician Leonhard Euler suggested the Earth was hollow:

> In the eighteen century Leonhard Euler, a Swiss mathematician, replaced the multiple spheres theory with a single hollow sphere which contained a sun 600 miles wide that provided heat and light for an advanced civilization that lived there.

However, in this case, the truth of what Euler actually suggested was different. Mathematician Ed Sandifer writes, in an April 2007 article titled "Euler and the Hollow Earth: Fact or Fiction?"[28], that in 1833, Euler was actually doing a thought experiment, in which he considered digging a hole to the centre of the Earth.

Sandifer explains that in Letter XLIX[29], Euler includes an illustration, Fig 30 (right) to show that depending where one is on the Earth, "down" is actually a different direction, so to speak. Euler, in considering how the force of gravity acts on a body, then considers digging a hole right through the middle of the Earth - as a thought experiment. Euler explains (referring to Fig 31) that, whether one falls from point A, B, E or F, one would always fall towards the point O - which is the centre of the Earth. Euler goes further in this line of thought in Letter L, writing:[30]

Fig. 30.

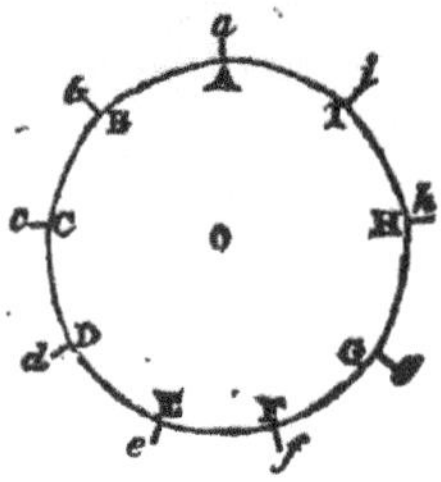

Fig. 31.

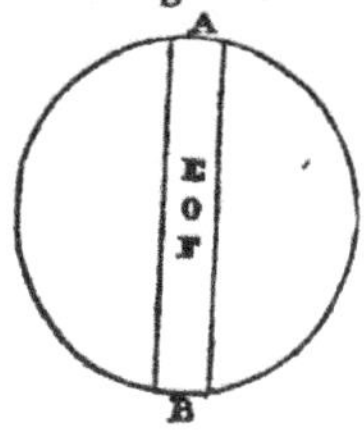

> Let us now return to the aperture made in the Earth through its centre; it is clear that a body at the very centre must entirely lose its gravity, as it could no longer move in any direction whatever, all those of gravity tending continually towards the centre of the Earth. Since, then, a body has no longer gravity at the centre of the Earth, it will follow, that in descending to this centre its gravity will be gradually diminished; and we accordingly conclude that a body penetrating into the bowels of the Earth loses its gravity, in proportion as it approaches the centre. You must be sensible, then, that neither the intensity nor the direction of gravity is a consequence from the nature of every body, as not only its intensity is variable, but likewise its direction, which, on passing to the antipodes, becomes quite contrary.

This observation will become important later on! Also, in Letter L, Euler talks about the acceleration due to gravity being different at the poles and the equator - which had been discovered by the time he wrote these letters in 1775.

Following in Chronological order, we can consider what Sir John Leslie, Scottish Physicist and Mathematician[31] wrote on page 450 of his 1829 book "Elements of natural philosophy" (2nd edition)[32]

> *The idea which I formerly threw out in the article Meteorology of the Supplement to the Encyclopaedia Britannica that the ocean may rest on a subaqueous bed of compressed air is therefore not devoid of probability.*

He included some calculations which suggested the Earth may not be just solid matter although he also considered the density of various materials when they were compressed in the depths below the Earth's surface:

> *If we calculate for a depth of 395$^{3}/_{5}$ miles, which is only the tenth part of the radius of the Earth, We shall find that Air would attain the enormous density of 101960 billions; While, at the same depth, Water would acquire but a density of 4.3492, and Marble only 3.8095.*

Also in the 19th Century, the idea of a Hollow Earth was popularised by an American Army Officer names John Cleves Symmes Jr. This was written up in some detail in Southern Bivouac, Volume VI[33]. In April 1818, Symmes made a public, written declaration, as shown below[34]:

LIGHT GIVES LIGHT, TO LIGHT DISCOVER—" AD INFINITUM."

ST. LOUIS, (Missouri Territory,)
North America, April 10, A. D. 1818.

TO ALL THE WORLD!

I declare the earth is hollow, and habitable within ; containing a number of solid concentrick spheres, one within the other, and that it is open at the poles 12 or 16 degrees ; I pledge my life in support of this truth, and am ready to explore the hollow, if the world will support and aid me in the undertaking.

Of Ohio, late Captain of Infantry.

N. B.—I have ready for the press, a Treatise on the principles of matter, wherein I show proofs of the above positions, account for various phenomena, and disclose *Doctor Darwin's Golden Secret.*

My terms, are the patronage of this and the new worlds.

I dedicate to my Wife and her ten Children.

I select *Doctor S. L. Mitchell*, *Sir H. Davy* and *Baron Alex. de Humboldt*, as my protectors.

I ask one hundred brave companions, well equipped, to start from Siberia in the fall season, with Reindeer and slays, on the ice of the frozen sea ; I engage we find warm and rich land, stocked with thrifty vegetables and animals if not men, on reaching one degree northward of latitude 82 ; we will return in the succeeding spring.   J. C. S.

It seems that it was primarily Symmes declaration (he never wrote a book) and lectures which triggered off a whole raft of other tales about a hidden land inside the Earth. As you can read above, Symmes even planned a trip to the North Pole to find the Polar openings! There is even a monument to Symmes in Hamilton, Ohio[35]!

In 1864, Jules Verne's novel "Journey to the Centre of the Earth," mentioned earlier, was published. An article by Joshua Glen on Hilobrow.com has further notes[36]:

> *Jules Verne's "Journey to the Centre of the Earth," a survey of mid-19th-century geological controversies thinly disguised as a ripping yarn set in a*

dinosaur-inhabited subterranean realm, was a best seller when it was first published in France in 1864. Five years later, an upstate New York alternative healer named Cyrus Teed had a vision revealing that the Earth is hollow and that we're all living on its concave inner surface. He eventually renamed himself Koresh, and established a thriving colony of hollow-earthers in the wilds of Florida. Then, in 1871, "The Coming Race," by Edward Bulwer-Lytton, in which the narrator discovers an advanced civilization underground, helped give rise to dozens of science-fiction novels in which travellers penetrated the icy polar realms and descended into a well-lighted netherworld via a whirlpool or tunnel.

Published by M.L. Sherman (M.D.) in 1875, a treatise entitled "The Hollow Globe" by Prof WM F. Lyon[37] describes more ideas about a hollow Earth. This seems to be a curious work - which is clearly not factual, but is not stated to be fiction either. In the preface[38] we can read:

*We do not claim that the teachings contained in this work are infallible, **neither are they presented in an authoritative manner.** But, we do claim, that it contains more original, natural and "startling ideas, which are of great interest to civilized humanity, and which seem to be entirely irrefutable, than any book of its size, that has made its appearance in modern times.*

The subject matter is described thus:

The central idea contained in the following work and the one that most of these chapters are designed to substantiate is, that this globe is constructed in the form of a hollow sphere, with a shell some thirty to forty miles in thickness, and that the interior surface which is a beautiful world in" a more highly developed condition than the exterior, is accessible by a circuitous and spirally formed aperture that may be found in the unexplored open Polar Sea, and this opening affords easy navigation, by a broad and deep channel leading from one surface to the other, and that the largest ships or steamers may sail or steam either way, with as much facility as they can pass through any other winding, or somewhat crooked channel.

It seems one can find several similar works which were published in the late 19th and early 20th century. Some of the ideas in these works are considered, by some "believers," to be truthful and accurate even today. We will explore some variants of the ideas/tales below.

Agharta (or Agartha), UFOs and Nazis

"Agharta" was the title of a 1964 book by Robert E Dickoff[39] (born in Germany in 1904), who was the founder of the American Buddhist Society and Fellowship (Inc)[40] . In his book, he describes Agharta as being "a vast underground terminal city" which is a "holy abode of the Buddhist world." This book seems to be confused with part of another book - Dr Raymond Bernard's book "The Hollow Earth" - also published in 1964[41]. The latter book makes a connection between Hollow Earth accounts and flying saucers/UFOs. It also discusses the alleged trips of Admiral Richard Byrd - which we will study in more detail in a later chapter. This work also refers to the story allegedly told to Willis George Emerson by a Norwegian Sailor Olaf

Jansen. We will also study this supposedly true story and attempt to verify some of the details.

What you will also find when reading other "Hollow Earth" literature are references to the Thule Society and the Nazis - and you will read stories about how Hitler may have sanctioned an expedition to the North Pole to find the entrance to the inner Earth. Similarly, when searching for "Hollow Earth" information on the web, an intricate diagram by Max Fyfield showing "Agharta" commonly appears. It is a diagram worthy of close inspection - as Fyfield has managed to pack many elements into the illustration - approximately one for each of the stories that are told about "Agharta"! You will find a fair number of postings on the web (some of which I have already referenced) giving much more detail about "Hollow Earth" stories, theories and so forth. Here, we will attempt to focus on some more serious research and we will attempt to verify some of the claims made and the accounts that can be read.

You will see several similar diagrams and most of these have 2 main features in common:

- There are holes at the North and South Pole
- There is a "Central Sun"

Although this diagram is mostly fantastical (judging by its style, Fyfield was a comic artist - but I have not been able to confirm this), it also shows an interesting element, which we will return to later. It shows that "the centre of gravity" is in the hollow Earth's shell - 400 miles below the surface.

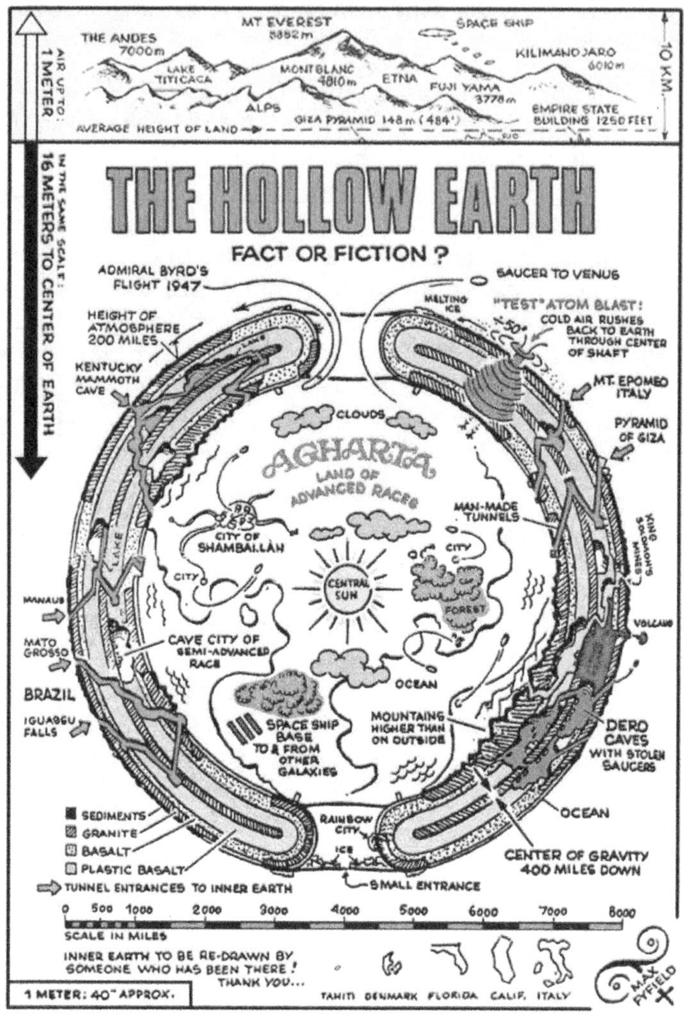

"Land of No Horizon"

In about 2005, an interesting book called "Land of No Horizon" by Kevin (an Art Teacher) and Matthew Taylor (a Web Designer) was published. There is a considerable amount of information about this book on an archived version of the website[42]. However, it appears that the site went offline sometime after 2013. Some years ago, I emailed the site, but did not receive a response. The domain www.hollowearththeory.com is now up for sale. The interesting diagram, below, is found on the archived version of their website.

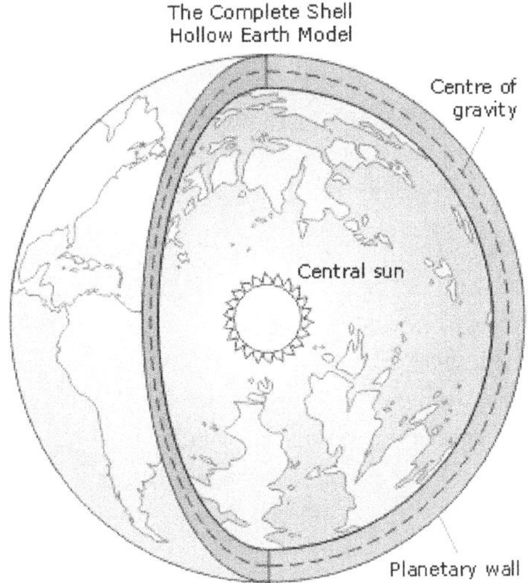

The Taylors were interviewed on the "Coast to Coast AM" show on 24 Nov 2005.[43] A synopsis of the show[44] includes a description relating to the diagram above:

Father and son researchers Kevin & Matthew Taylor presented their … idea that there is an inner sun and a "perfect environment" for supporting life inside the Earth. In fact, the Taylors contend that beings, thousands of years ahead of us evolved there, and orchestrated humanity's development. "I believe UFOs, gods and Inner Earth people are all one thing," said Kevin.

- *Because there is little to no gravity, trees and other life forms grow much taller or bigger.*
- *There is constant warmth and daylight from the radiating sun.*
- *A shallow ocean travels through the land masses.*
- *The visual perspective is much different than here as there is no horizon. Continents appear to float in the clouds.*

Planets the size of Earth and Mars evolve to become hollow and then expand under their own gravitational power. They also related that the Great Flood from the time of Noah was created by the Hollow Earth inhabitants who released the water in a careful manner in order to correct an imbalance and guard against a potential meteor impact.

In the broadcast, the authors explain that the title of their book is derived from the notion that if you were walking on the inside surface of the hollow Earth, the land in front of you would actually rise up - and into the clouds (assuming there were clouds and an atmosphere in there).

The interview was a bit of a mixed bag, as the authors discussed possible entrances to the inner Earth being in the Bermuda triangles and other places. They suggest there are entrances/channels which run between the inner and outer world. It was through these channels that water was released in the "great flood."

More interestingly, and perhaps a little more realistically, they also suggested there is warm water coming out of these channels which are near the poles and it was these warm currents that were causing some anomalous ice melts in some places.

The authors also suggested that the reason that the bottom of Izu-Bonin trench[45] (east of the Japanese islands) is rising is because it is a channel or opening "through to the inner world."

They include descriptions of the Earth's interior, but this is entirely speculative. The story they describe doesn't seem to be always consistent in its details, though I have not been able to read their book to see if these apparent inconsistencies are explained. For example, they talk about a land being on the inside but don't mention where the water is in the inner world (i.e. in the 2005 interview, they don't mention any oceans or large bodies of water in the inner world which might be gushing through into the outer world.)

There are quite a few places in the interview where they "hand wave" and don't seem to discuss much of the actual evidence they say they have found.

Further details about their theory are found on their (archived) website (I have corrected a couple of small spelling errors):

> Based on a combination of the **Hollow Earth and expanding Earth theories**, it provides an alternate explanation for the drifting continents phenomenon thus making the tired **Plate Tectonics theory obsolete.**
>
> Based on our current understanding of gravity The Land of No Horizon shows how the accumulation of matter during the planet forming process naturally produces a planet structured differently to what is currently theorised. **It is also shown how a planet hollows out and expands under its own gravitational power.**
>
> The hollow planet structure can explain many mysteries that have plagued us for centuries such as unusual Impact crater characteristics on terrestrial planets, the mysterious Red Spot on Jupiter and seismic wave data from earthquakes here on Earth. Understanding outgassing and atmosphere formation on a hollow planet model helps us explain past mysteries such as the great flood on Earth and the Floods on Mars.
>
> An expanding Earth provides the driving force behind the drifting continents, mountain building and earthquakes and is also accountable for changing the value of gravity over time. In the past when the Earth was smaller centrifugal force from a much faster speed of rotation reduced the effects of gravity in equatorial regions. **This reduction of gravity is what allowed the great dinosaurs and all past life to grow to much larger sizes.**

The points that I have emboldened, particularly relating to Earth expansion, will be revisited later in considerable detail, though we will be looking at those issues from a somewhat different perspective than the Taylors did. Further descriptions on the archived website show that the Taylors share an interest in the same sorts of topics as me - as they also discuss gravity and human origins in their book - and how their Hollow Earth theory is related to those issues. (I touched on the topic of human origins in my previous book, "Acknowledged."[46])

3. Olaf Jansen and "The Smoky God"

"The Smoky God"[47] was written by Willis George Emerson (a novelist) and published in 1908 by Forbes and co. in Chicago. It is presented as a true story, and I remember first hearing about it in a radio interview - probably in 2011 or 2012. The person being interviewed sounded sufficiently convinced that the Olaf Jansen account was true that I decided to read it and try to verify it. The story includes some specific dates and locations, which means that there is more opportunity to check its plausibility.

On 3rd April 1829, Age 19, Jansen and his father set out from Stockholm, passing Gotland (Gothland) Island and Oland (Oeland) Island. They arrived in Spitzbergen on 23rd June - early in the Arctic summer. They arrived in Franz Josef Land on about 30th June. This meant that in about 10 weeks, they sailed about 3500 miles (approx. 50 miles per day) - this seems reasonable too.

The story told of "dodging Icebergs and almost running out of Food and Water." After leaving Franz Josef Land, they claim to have spent 1 day on an Island North of there, but this does not seem to appear on any map. Leaving this unknown island, they ended up in "the Land Beyond the North Pole"…

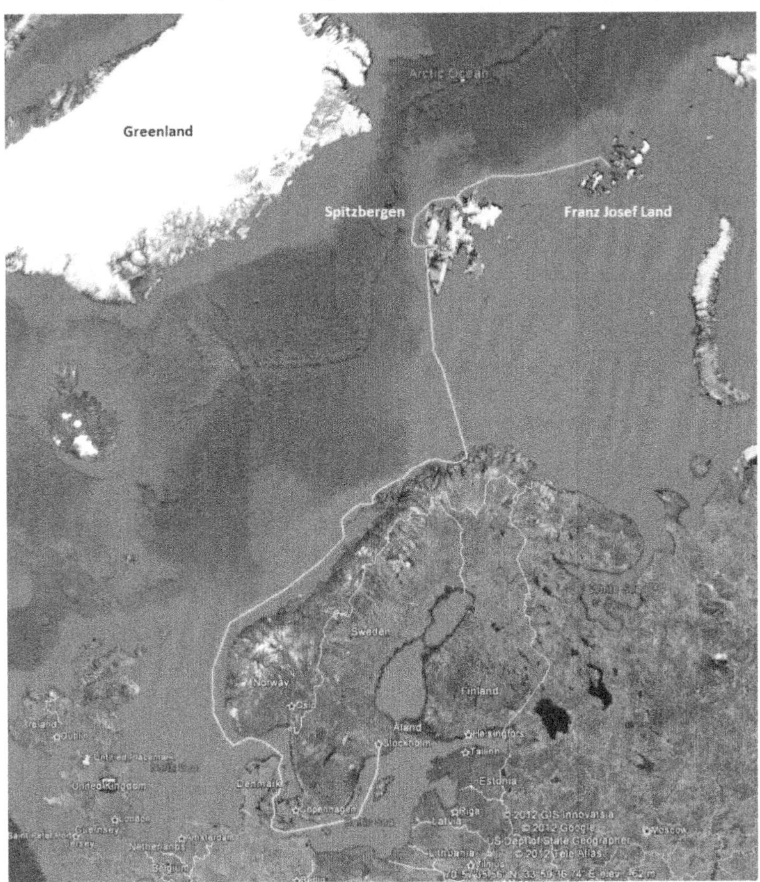

Tracing this journey out on a google map (see above), shows that it could have taken place as described - the locations and timings seem to be realistic. The only obvious problem is the "missing island" - and it does not seem that they would have travelled across an active volcanic fault line, where we can sometimes find that new islands sporadically appear[48]!

...could hardly be said to resemble the sun except in its circular shape

Sailing North from the unknown island, they later entered a freshwater sea and could see a different "kind of sunlight" in the sky. They don't report getting "sucked down" into a hole or a vortex or anything like that. In the story, Jansen reported:

> *"For several days, when I looked for this star, it was always there directly above us."*

It is unclear how this would be visible when, in June, at or near the North Pole, the Sun is visible most of the time. Some might argue they were seeing some effect that is only visible near the supposed holes in the Earth's polar regions.

After more days of sailing, they come across another land and they are greeted by "giants."

Jansen writes:

> *The immense craft paused, and almost immediately a boat was lowered and six men of gigantic stature rowed to our little fishing-sloop. They spoke to us in a strange language. We knew from their manner, however, that they were not unfriendly. They talked a great deal among themselves, and one of them laughed immoderately, as though in finding us a queer discovery had been made. One of them spied our compass, and it seemed to interest them more than any other part of our sloop.*
>
> *Finally, the leader motioned as if to ask whether we were willing to leave our craft to go on board their ship. "What say you, my son?" asked my father. "They cannot do any more than kill us."*
>
> *"They seem to be kindly disposed," I replied, "although what terrible giants! They must be the select six of the kingdom's crack regiment. Just look at their great size."*
>
> *"We may as well go willingly as be taken by force," said my father, smiling, "for they are certainly able to capture us." Thereupon he made known, by signs, that we were ready to accompany them.*

Jansen then recounts how they toured the land of the Inner Earth for about 2 years. This part of the story holds some interesting details:

> *...we were taken overland to the city of "Eden," in a conveyance different from anything we have in Europe or America. This vehicle was doubtless*

some electrical contrivance. It was noiseless, and ran on a single iron rail in perfect balance. The trip was made at a very high rate of speed. We were carried up hills and down dales, across valleys and again along the sides of steep mountains, without any apparent attempt having been made to level the Earth as we do for railroad tracks. The car seats were huge yet comfortable affairs, and very high above the floor of the car. On the top of each car were high geared fly wheels lying on their sides, which were so automatically adjusted that, as the speed of the car increased, the high speed of these fly wheels geometrically increased.

Jules Galdea explained to us that these revolving fan-like wheels on top of **the cars destroyed atmospheric pressure, or what is generally understood by the term gravitation, and with this force thus destroyed or rendered nugatory, the car is as safe from falling to one side or to other from the single rail track as if it were in a vacuum**; *the fly wheels in their rapid revolutions destroying effectually the so-called power of gravitation, or the force of atmospheric pressure or whatever potent influence it may be that causes all unsupported things to fall downward to the Earth's surface or to the nearest point of resistance.*

I just found the highlighted description interesting, because of my other research into antigravity devices[8]!

Perhaps it is all too much like the Jules Verne story. The hardest part to accept, for me, was that Jansen claimed to have entered through the north polar hole and exited (2 years later) through the south polar hole! This would be quite a distance to travel (even in 2 years!)

Let us now consider the part where he recounts when he was rescued, when they were once again at the mercy of the oceans and the weather. Jansen explains that their boat was destroyed and his father was lost with the boat. Jansen managed to survive by clinging to an iceberg:

Then the hand of the Deliverer was extended, and the death-like stillness of a solitude rapidly becoming unbearable was suddenly broken by the firing of a signal-gun. I looked up in startled amazement, when, I saw, less than a half-mile away, a whaling-vessel bearing down toward me with her sail full set.

Evidently my continued activity on the iceberg had attracted their attention. On drawing near, they put out a boat, and, descending cautiously to the water's edge, I was rescued, and a little later lifted on board the whaling-ship.

I found it was a Scotch whaler, "The Arlington." She had cleared from Dundee in September, *and started immediately for the Antarctic, in search of whales. The captain, Angus MacPherson, seemed kindly disposed, but in matters of discipline, as I soon learned, possessed of an iron will.*

The captain sent for me and again questioned me concerning where I had come from, and how I came to be alone on an iceberg in the far-off Antarctic Ocean. I replied that I had just come from the "inside" of the Earth, and proceeded to tell him how my father and myself had gone in by way of Spitzbergen, and come out by way of the South Pole country, whereupon I was put in irons. I afterward heard the captain tell the mate that I was as

Jansen then made it back to Stockholm, but following his return, he claims that he was incarcerated in a mental institution for 28 years - because he told this fantastic tale to his Uncle, who then decided he must have gone mad (perhaps because Jansen's father was killed late in the voyage and this had obviously taken its toll on Jansen's sanity). In later life, Jansen travelled to the USA and in the story, he accurately records the date of the 2nd inauguration of President McKinley as March 4th, 1901.

It appears that Jansen did not write his story down in full until after the 28 years of horror he claimed to have suffered. He was unwilling to tell his tale in full until near his death - and he did so, to Willis Emerson - the novelist. So it appears Jansen never profited from his fantastic tale.

Checking Olaf Jansen's Amazing Story Further

Although we have already noted that Jansen's alleged route and trip to the north pole seems plausible, and he correctly noted the 1901 presidential inauguration date, this is not enough to give his story credibility. It's true to say that the myths and stories regarding the "Hollow Earth" had already been in circulation for *decades* by 1829, and that it was later in the 19th century that the tales became more popular.

One detail that is given in Olaf Jansen's account that is checkable is the reference to the Dundee Whaling Ship called the Arlington. If you consider the dates given, this ship should have left Dundee in September 1830. In my efforts to check this reference, I found that a chap called Malcolm Archibald has written a book about the Scottish Whaling Industry[49], so I wrote to ask him about the use of Scottish Whaling ships in the *Antarctic* in September 1830.

He kindly replied:

based around South Georgia, the Falklands and sometimes Staten Island off Cape Horn] but they were few and far between and concentrated on seals. It was much easier for the South Sea Whalers to hunt in the South Atlantic and the Indian Ocean. I can only think of one Scottish South Sea whaler and that was Captain Weddell in Jane of Leith: he sailed south in 1822 and, having read his journal, he made no mention of picking up a castaway.

The name Angus McPherson I would also think was a 'stock' name for a Scottish sea master: part of the tartan and shortbread image used by people with limited or no knowledge of the country."

Malcolm could find no record of a whaler named "Arlington."

Conclusion

It is more than likely that the Olaf Jansen story is just "a good yarn" - written to cash in on the interest that people had in "fantastic tales" of a "hollow Earth." With the early 1900s popularity of novels by Jules Verne such as "The Underground City" (1877) and the afore-mentioned "Journey to the Centre of the Earth," a publisher would likely be quite happy to publish the "Smoky God" story, which had a similar setting and plot.

Although some details of the story seem to be plausible, there aren't enough elements that can be checked and those that can be checked don't seem to be correct - such as a missing island and a whaling ship that could not have been there. Nevertheless, the story, which is about 16,000 words long, makes for an enjoyable and entertaining read.

4. Holes and Bases at the Poles?

One of the most obvious problems with the "hollow Earth" idea/theory is the often discussed "Polar Holes." In the literature, it is quite common to see illustrations like those shown below.

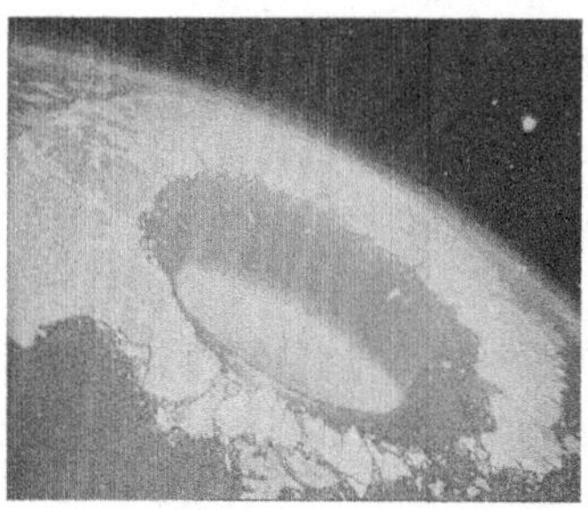

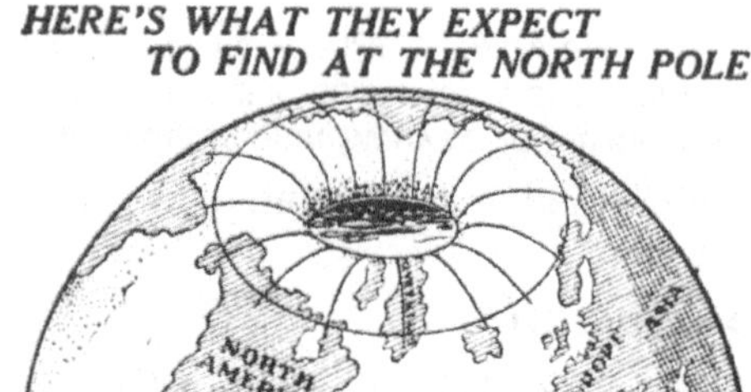

However, let us pause for a moment and point out that the presence or absence of polar holes does not settle the matter of what is at the Earth's core - it is merely one aspect of a theory which, we will show, has no basis in our physical reality.

The wording I used is deliberate - because I personally think that we sometimes interact and experience a "non-physical reality" - and perhaps there is some sort of "polar hole" aspect that we might consider to be associated with a "non-physical reality." Indeed, if we consider the path of lines of magnetic force into and out of the Earth, might we consider that these lines of force are travelling through some type of portal or hole?

Location and Size of Physical Polar Holes

If you read the various articles and postings, you will find many different illustrations and within those, different sizes for the Polar Holes are stated or implied. Locations are rarely given, but on one webpage[50] (since removed from the original URL, though the site is still active), I found the following information regarding an alleged north polar hole:

On a map, the perimeter would begin at about 784 miles from Point Barrow, Alaska and the centre of the polar opening would be located at 87.7 N Latitude and 142.2 E Longitude (2.3 degrees from the pole). From Point Barrow, Alaska, the perimeter would begin at 784 statute miles and the inner continent, which I estimate to be located halfway through the polar opening, at 1,413 miles.

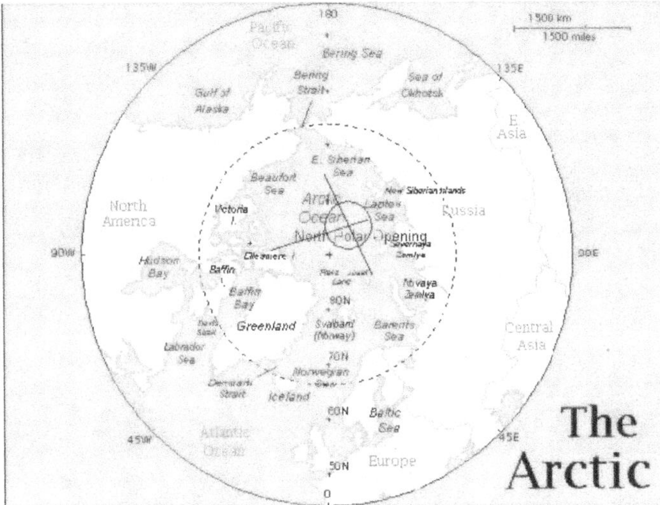

The map/image above is one of the few which gives a specific location for a north polar hole. Mostly, you will find vague descriptions and "woolly accounts" or fanciful pictures with, perhaps the exception of…

Satellite Images Show Holes!

Again, when searching for information on the Hollow Earth, you will quickly come across one or both of the images below. Surely, as we have 2 images, this must mean the holes are really there!

This photograph was taken on 6 January 1967, by the satellite ESSA-3, clearly showing the hole at the North Pole.

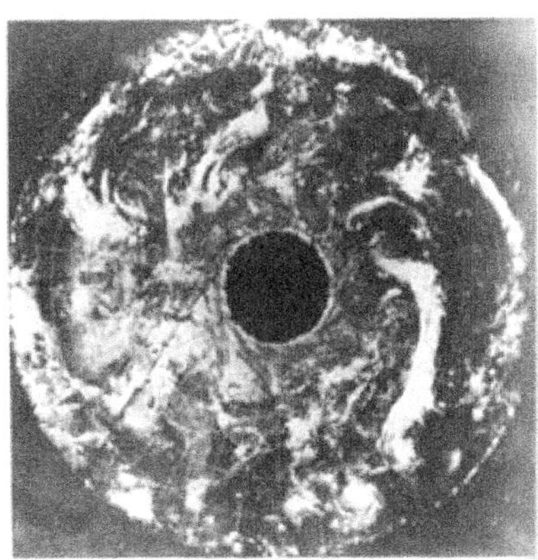

The remarkable photograph taken by the ESSA-7 satellite on 23 November 1968. There is almost no cloud cover; the ice fields on the surface can be observed, and the hole at the North Pole can be clearly seen.

We will study these images in more detail below.

Some people have tried to say there aren't any photos of the North Pole from Space… However, these are from the European Space Agency (ESA) website, but we can also see holes!

Arctic ice concentration in 2005[51]

So, surely this is evidence of an entrance there to the inner Earth!

However, we need to be careful here - and make sure we know what the images are really showing…As Sean Ellis observes on his "Mote Prime" Website[52] regarding the first two 1967/8 North Pole Satellite Images, above:

The first photograph referred to by McElwaine was taken by ESSA-3 on January 6, 1967:

The caption includes the words "clearly showing the hole at the North Pole." If this is indeed the case, then it is nothing short of remarkable, as the North Pole is in complete darkness at this time of year!

There are other extremely suspicious things about this photograph. Not just the "hole", but the entire visible hemisphere of the Earth is in sunlight - clearly impossible unless the pole is oriented directly toward the sun. This is not, of course, the case. At no time during January 1967 was the entire Northern Hemisphere simultaneously bathed in sunlight.

Oh, and another thing. If the interior of the Earth is illuminated by "a glowing ball of plasma, about 600 miles in diameter", why don't we see a bright disk instead of a dark hole?

Answer: it's a composite photo. ESSA-3 was a polar-orbiting satellite, and this image is a clearly a montage produced by NASA from a number of separate half-lit images to give a measure of global cloud cover.

Ellis continues:

The third (left hand ESA) image is clearly a composite - you can see the image strips. The fourth (coloured ESA) image is a false colour image - not showing true optical information. None of the four images include any data for the North Pole itself… Yet, you will still see these sorts of images (particularly the first two) listed as being proof that polar holes have been observed.

True North Pole Images

Satellite images taken above the north pole are rare, due to the difficulty/expense of putting a satellite into a polar orbit (using white world technology, that is). However, there are some "non-composite" images of the north pole for example:

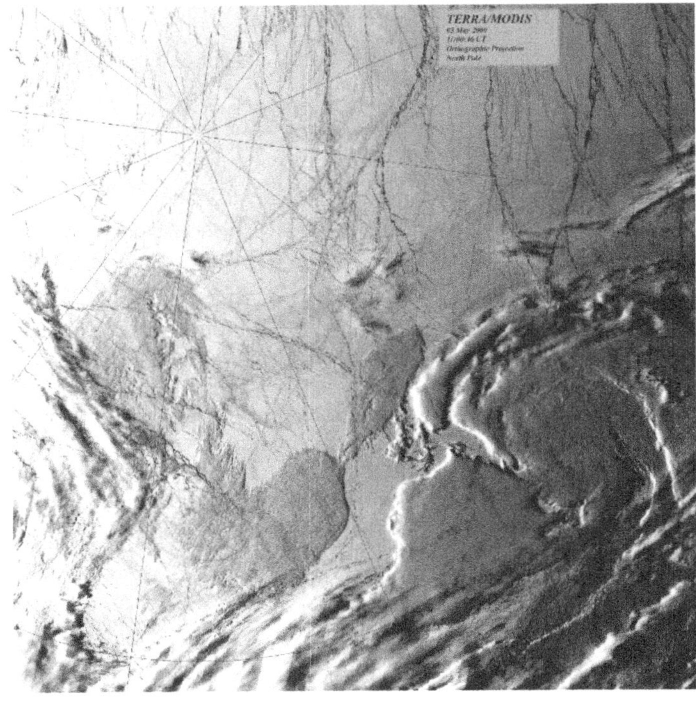

Credit: Image by Allen Lunsford, NASA GSFC Direct Readout Laboratory; Data courtesy Tromso receiving station, Svalbard, Norway[53]

This true-colour image over the North Pole was acquired by the MODerate-resolution Imaging Spectroradiometer (MODIS), flying aboard the Terra spacecraft, on May 5, 2000. The scene was received and processed by Norway's MODIS Direct Broadcast data receiving station, located in Svalbard, within seconds of photons hitting the sensor's detectors.

In this image, the sea ice appears white and areas of open water, or recently refrozen sea surface, appear black. The irregular whitish shapes toward the bottom of the image are clouds, which are often difficult to distinguish from the white Arctic surface. Notice the considerable number of cracks, or "leads," in the ice that appear as dark networks of lines.

Throughout the region within the Arctic Circle leads are continually opening and closing due to the direction and intensity of shifting wind and ocean currents. Leads are particularly common during the summer, when temperatures are higher and the ice is thinner. In this image, each pixel is one square kilometre.

Such true-colour views of the North Pole are quite rare, as most of the time much of the region within the Arctic Circle is cloaked in clouds.

Despite images like the ones above, sometimes you'll read about "no-fly" zones over the poles…[54]

Since no polar orbiting satellite can go over the polar opening, the only conclusion is that this no-fly zone for polar orbiting satellites is where the polar opening is located. It is an area about 500 miles centred over the pole down and directly over Northland, Russia.

So, it does go to show that some people will believe a good yarn over evidence…

North Pole - "No Fly" Zone?

People ask if planes can fly over the poles. Rumours sometimes circulate that pilots are "forbidden" from flying over the north pole. Apparently, it may be somewhat discouraged due to "clear air" directives. However, when I was originally researching this topic, I got in touch with a chap called Don Daniels who, for several years, was a captain with the US United Airlines company. I asked him about flying over the north pole. He wrote back:

We are not allowed to fly directly over the North Pole, the stated reason being that the older navigation systems on, for instance, the B-747 would get temporarily confused as to which South to reference to. Newer navigation systems like on the B-777 mathematically switch the poles to the equator when we are far north or south and thus there is no ambiguity. However, they tell us to not even ask as we will not get permission to deviate directly over the pole, and it would be considered a gross navigation error.

I have been within 60 miles of the North Pole on a number of occasions, and while it is surprisingly smoggy or hazy, I should have been able to see an opening of any significant size.

He explained it was in 2006 - when he flew near the North Pole and he sent me some photos! (See below). Don Daniels, who was also involved with Dr Steven Greer's disclosure efforts, shares an interest in many of the topics I've written about and he too has written a book called "Evolution Through Contact."[55]

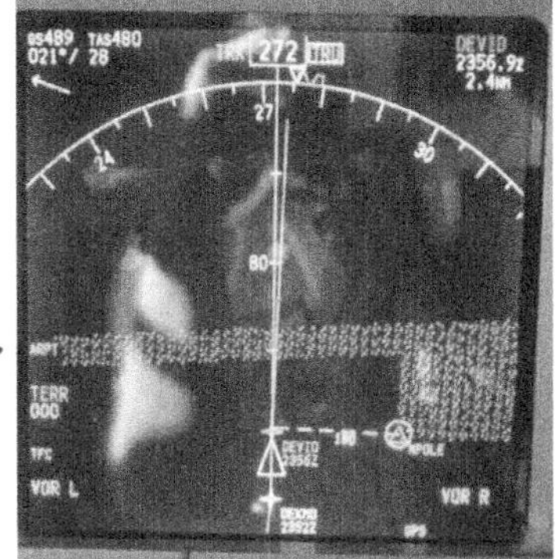

Flying Near the North Pole! (The view that Don Daniels had in 2006)

Michael Palin at the North Pole

Many of the sources that want to talk about "holes at the poles" seem to completely ignore that people have been to those same poles. In May 1991, Michael Palin, Monty Python - author, writer and actor turned explorer – visited the North Pole - which is on a shifting ice field. He then set off on a 6-month journey towards the South Pole. He recounts this in the video[56] and book[57] "Pole to Pole."

Michael Palin - South Pole in
May 1991

I'm squeezed tight into a small, noisy aeroplane descending through stale grey cloud towards an enormous expanse of cracked and drifting ice. With me are Nigel Meakin and his camera, Fraser Barber and his tape-recorder and Roger Mills and his pipe. With our two pilots, Russ Bomberry and Dan Parnham, we are the only human beings within 500 miles. Outside my window one of our two propeller-driven engines slowly eats away at a fuel supply which must last us another six hours at least. In little more than ten minutes our pilot will have to fashion a landing strip out of nothing more than a piece of ice - strong enough to withstand an impact of 12,500 lbs at eighty miles an hour.

You can read the rest of Palin's account on the website linked above.

A South Polar Hole - Or a Secret Polar Base?

Depending, again, on which web pages you read, you will be told that either there is an entrance at the South Pole to the inner Earth, or there is a secret Nazi Base or other base there… But what does the evidence show?

It shows that there is indeed a base at the South Pole! It is called the Amundsen-Scott Station and it is run by the National Science Foundation[58].

Apparently, Michael Palin was somewhat surprised at the scale of what he found when he went there on 4 December 1991.

In his book[59], and in a video[60], Palin describes his experiences of landing at the South Pole.

12.30 a.m. Over the noise of the engine Dan shouts back that we are forty-seven minutes from the Pole.

1.00 a.m. Radio communication from air traffic control at the South Pole base.

'There is no designated runway and the US government cannot authorize you to land. How do you copy?'

Dan: 'OK.'

'OK. Have a good landing.'

Scott gives Rudy a shot of oxygen. The effects of the height can now be clearly felt. Shortage of breath, every movement requiring twice the effort.

1.10 a.m. We can see the South Pole ahead. It is somewhere in the middle of a complex of buildings dominated by a 150-foot-wide geodesic dome. Vehicles and building materials are scattered about the site. It is the busiest place we've seen in Antarctica.

1.20 a.m. We land at the Amundsen-Scott South Pole Station, scudding to a halt on a wide, cleared snow runway.

Two well-wrapped figures from the base wait for us to emerge from the plane, and shake our hands in welcome, but the senior of them, an American called Gary, advises us that it is not the policy of the National Science Foundation, who run the base, to offer assistance of a material kind to NGAs - Non-Government Agencies - such as ourselves. Scott confirms that our expedition is self-sufficient and that Adventure Network has a cache of fuel and accommodation located nearby.

Michael Palin at the South Pole in December 1991

In a Coast to Coast Broadcast, David Hatcher Childress reported that Palin had been told there were 7 underground levels at the South Polar base, but Palin does not mention this in his book and I have found no other evidence to support this claim.

The original base was built in the early 1970s[61]:

...from 1975 to 2008, home at the Pole was known familiarly as 'the Dome,' or, affectionately, 'Dome, Sweet Dome.' Built from 1971 to 1975, the former base was a silver-grey aluminium geodesic dome, 50m in diameter at its base and 15m high. It covered three structures, each two stories high, which provided accommodations, dining, laboratory and recreational facilities.

By 2008, the dome had become unsafe, due to repeated coverings of snow and ice and it was decided to build a new station, which is elevated and has an aerodynamic shape to reduce the snow build up. The old base was dismantled.

Jan Lamprecht and Polar Holes

Later we will refer to Jan Lamprecht's research on seismic waves as they relate to the core of the Earth. Here, however, it is relevant to mention his book[62] "Hollow Planets: A Feasibility Study of Possible Hollow Worlds."[63] In this book, he spends a considerable amount of time discussing the idea of polar holes, but his discussion is largely focused around anecdotes and what these anecdotes might mean. He also considers phenomena such as the polar auroras and the ozone hole. At no point, however, does he seem to address the points I raised above. This is a shame, because his separate article on the reinterpretation of the seismological studies of the Earth (discussed later) seems much more relevant and cogent to what we will discuss in later chapters of this book.

Summary

In this chapter I have tried to find more direct evidence that there may be holes in the poles, leading to an "inner Earth" - however, I can find no such evidence.

5. Admiral Byrd and Polar Exploration

Another oft-quoted story about a journey into the Hollow or Inner Earth is that of Admiral Byrd. During his expeditions to the South Pole, Byrd allegedly discovered a "secret land." There are various versions of this story and I have, in the past, had several people writing to me to ask me about them or suggest that I should "check them out." For example, a posting has appeared quite near the time of writing of this chapter (June 2019), where we can read[64] about "Admiral Byrd's Hollow Earth Story":

> *In 1947 Byrd allegedly undertook another flight over the North Pole. This time, in a diary that didn't surface until years later, Byrd recorded events as he flew into the Earth, and encountered the inhabitants who live there.*
>
> *Allegedly, he landed his craft, and spoke with a representative of the city of Agartha (referred to as Ariana in some accounts), who reprimanded him for humanity's recent invention of the atomic bomb, and warned that a dark age is to come if humans don't shape up. Byrd was sent on his way, with instructions to bring this message back to the surface people.*
>
> *Byrd was a decorated military man and public figure. It's seems unlikely that he would lie about something he experienced on an expedition, but quite likely he would want to keep it a secret. There are few explorers who would be considered more reliable than Richard E. Byrd, so if Byrd did keep such a diary it probably happened.*

So what is the actual truth behind this story, and can we establish why this story appeared? Please read on…

Admiral Richard E Byrd

Before we go any further, we can of course state that Admiral Richard E Byrd (1888 - 1957) was indeed a real officer in the US Navy and he did indeed complete trips to both the north and south polar regions. Here we can reflect on how poorly researched articles like the one quoted above don't even bother to reference more reliable sources like Britannica.com, where one can find details of Byrd's Polar Expeditions[65]…

A Britannica article about Admiral Byrd notes:

> *On May 9, 1926, Byrd, acting as navigator, and Floyd Bennett as pilot made what they claimed to be the first airplane journey over the North Pole, flying from King's Bay, Spitsbergen, Norway, to the Pole and back….*

The article then discusses the possibility that Byrd didn't actually reach the pole, due to an oil leak, but this information didn't surface until 1996, when

Byrd's diary (or diaries) were revealed. (We will re-visit this element of the story later). Only 2 years after Byrd's North Pole Expedition, he was at the other end of the world…

Byrd's first Antarctic expedition (1928–30), the largest and best-equipped that had ever set out for that continent, sailed south in October 1928. A substantial and well-supplied base, called Little America, was built on the face of the Ross Ice Shelf…

The Britannica article also notes the funding of this 1928 expedition was given by John D Rockefeller and others. Byrd made at least two further trips to Antarctica:

In 1933–35 a second Byrd expedition visited Little America with the aim of mapping and claiming land around the Pole…

At the request of President Franklin D. Roosevelt, Byrd took command of the U.S. Antarctic service and led a third expedition to Antarctica in 1939–41, this one financed and sponsored by the U.S. government…

After World War II Byrd was placed in charge of the U.S. Navy's Operation High Jump. This Antarctic expedition, his fourth, was the largest and most ambitious exploration of that continent yet attempted and involved 4,700 men, 13 ships (including an aircraft carrier), and 25 airplanes. Operation High Jump's ship- and land-based aircraft mapped and photographed some 537,000 square miles (1,390,000 square km) of the Antarctic coastline and interior, much of it never seen before…

It is this last "Operation High jump" Antarctic Expedition which has been the one most commonly associated with Byrd meeting other beings or finding an entrance to the "Hollow Earth" etc. We will explore the possible reasons for this below. We can also note here that, perhaps unsurprisingly, Byrd was a freemason[66].

Operation High Jump

In 1947, the US launched a well-equipped expedition to Antarctica to follow up on earlier expeditions. We can get a useful overall view of what happened on this expedition if we watch a 90-minute film called "The Secret Land" (1948)[67]. David Glagovsky has written (on IMDB) an excellent synopsis of the film[68]:

The Secret Land - This film documents the largest expedition ever undertaken to explore Antarctica. The expedition, code named "Operation High Jump," was made by the U.S. Navy and involved 13 ships (including one submarine), 23 aircraft, and about 4700 men. The film was shot by photographers from all branches of the U.S. military. One purpose of the expedition was to explore and photograph several thousand square miles of inland and coastal areas that had not been previously mapped. Additionally, military planners wanted to evaluate whether military troops could successfully perform against an adversary in such an environment.

The film starts with a short speech by Secretary of Defence James Forrestal (who became quite well known in some stories connected to the aftermath of the Roswell case)[69]. Watching the rest of this remarkable (colour film) one gets a sense of the amount of skill, courage and resources that made the expedition very successful. For example, one memory I have of watching the film is when they discovered that the aircraft carrier they were using was too short for the larger planes to take off successfully. The solution? Well, that was obvious… they just needed to attach "jet assist bottles" to each plane and fire these when setting off, then drop them into the sea once sufficient speed had been attained[70]! All this said, it also seems to me like the film did have some "post-production" with some segments re-shot in a mock-up plane and edited in, to make the final film. (I could be wrong, but on reviewing the film, some shots seem like they were taken in a stationery plane and particular camera angles were chosen for certain shots.)

The Ice-Free Region in the Antarctic - Bunger Hills

On the web, much seems to be made of the ice-free area of land that was discovered by Byrd's expedition. The region was named "Bunger Hills,"[71] after the High jump pilot who first flew over it. This is shown in the "Secret Land" film in a segment where they fly over "300 square miles without snow." Bunger landed his plane on one of several freshwater lakes in the area. The temperature of the water was reported to be 38F (3C).

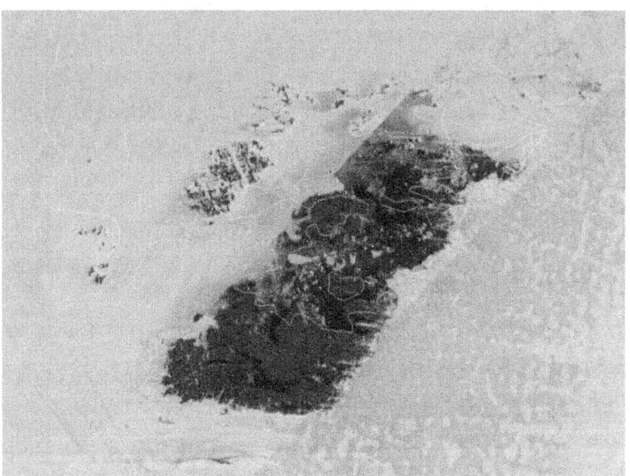

Aerial view of part of Bunger Hills Antarctica

In this same area, the Oazis Station[72] was built then opened by the Russians in 1956. The Station was later handed over to the Academy of Sciences of the Polish People's Republic when it was renamed the "A. B. Dobrowolski Station." It had limited usage up until the 1990s. It's not in use now though the buildings are still standing.

Ross Island and Mount Erebus

These are also near the continent of Antarctica and Mount Erebus is an active Volcano. It is more regularly active than most other volcanos in the world - and is also the most southerly.

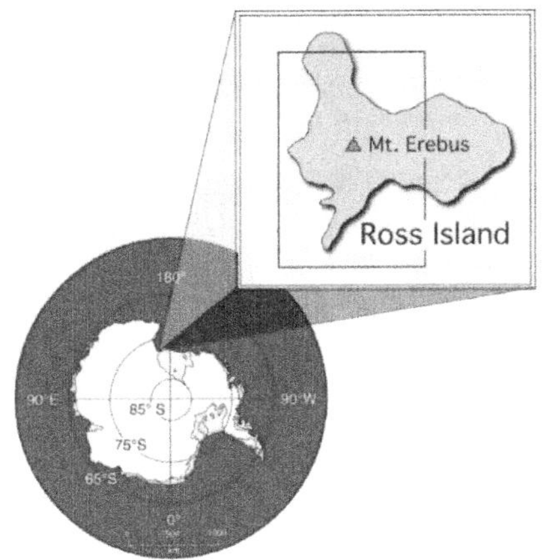

Location of Ross Island and the Volcano Mount Erebus in relation to Antarctica

The Island and Volcano were discovered in 1841 by polar explorer Sir James Clark Ross - the volcano was erupting at the time of its discovery. Ernest Shackleton climbed Mt. Erebus in 1907. An Atlas Obscura page[73] notes:

> *Because of the gas, the ice caves stay a consistent 32 degrees, making them a likely spot for undiscovered extremophiles. The volcanic gases heat their way through these ice caves and escape into the air to form enormous 60-foot chimneys of ice, or "fumaroles" with deadly volcanic gases pouring out from their tips.*

The Mythical Stories About Admiral Byrd

It seems very likely to me that the account of Bunger landing his plane on the (relatively warm) Fresh Water lake, and the account of seeing an active volcano have been conflated and/or confused with the idea that Byrd's plane somehow entered the inner Earth and discovered warmer conditions there. We have already seen how Britannica discussed Byrd's diary and how the North Pole over-flight might never have happened - but now let's revisit the diary and try to find other clues as to where the "Hollow/Inner Earth" stories came from.

The "Missing" Diary of Admiral Byrd

As noted earlier, Admiral Byrd made at least 3 Polar flyover trips/attempts before the 1946-7 Operation High jump. His trips took place in 1926 (North) and 1928-30 and in 1933-35. As can be seen by watching the "Secret Land" film, the 1946-7 Operation High jump was a large, very well-equipped expedition which made front-page headlines around the world. Reports about Byrd's movements were broadcast around the world daily from January 2nd 1947 until his return to America on April 14th 1947.

However, some of the stories talk of Byrd travelling to the *North* pole in *February* 1947. An article by Dennis Crenshaw[74], originally posted on his website "Hollow Earth Insider,"[75] looks at where the story of Byrd's 1947 *North* Polar trip came from. Crenshaw writes:

The first mention of a 1947 North Polar flight by the Admiral can be found in the book "Worlds beyond the Poles: Physical Continuity of the Universe" (1959) by a controversial self-proclaimed scientist, F. Amadeo Giannini. On page 13 of his book under the heading "The Changing Scene 1927-1947" he presents a list of things that happened during those years to support his theory. One of the entries is:

"1947: February "I'd like to see that land beyond the pole. The area beyond the pole is the centre of the great unknown." - Rear Admiral Richard E. Byrd before his seven-hour flight over land beyond the North Pole."

In 1928 Admiral Byrd published his account of the 1926 North polar flight in a book titled "Skyward." On page 196 Byrd reveals the following information:

"When our calculations showed us to be about an hour from the Pole, I noticed through the cabin window a bad leak in the oil tank of the starboard motor. When I took the wheel again I kept my eyes glued on the oil leak and the oil pressure indicator."

Compare this information to the following passage from the so-called "secret diary."

(Page 2) "--- Hours: Slight oil leak in starboard engine, oil pressure indicator seems normal, however."

Crenshaw rightfully considers the incredible odds of Admiral Byrd having the exact same problem with the same engine on two different flights, 11 years apart. Crenshaw also discusses other alleged statements made by Byrd, such as those contained in "The Secret Diary of Admiral Richard Evelyn Byrd."[76] This work describes a meeting between The Admiral and the "Master" of the Arianni or "Ari Anni." The Master tells Byrd there "is a dark time coming" but that:

"Yes, my son," replied the Master, "the dark ages that will come for your race will cover the Earth like a pall, but I believe that some of your race will live through the storm, beyond that I cannot say. We see a great distance… a new world stirring from the ruins of your race, seeking its lost and legendary treasures, and they will be here my son, safe in our keeping…"

Crenshaw explains that he mentioned his "Byrd" research to his friend Robert Van Aulen (an artist), who then sent him a video of the 1937 MGM motion picture "Lost Horizon." Van Aulen pointed out a scene in the film where the star, Ronald Coleman, has an audience with the Dali Lama in Shangra-La, a lost city in Tibet. The Dali Lama says:

"You, my son, (said the Master), "will live through the storm. You will preserve the fragrance of our history and add to it a touch of your own mind. Beyond that, my vision weakens … but I see in the great distance a new world starting in the ruins … But in hopefulness, seeking it's lost and legendary treasures, and they will all be here, my son, hidden behind the mountains under the blue moon, preserved as if by a miracle…"

So it seems like the "Byrd" story was plagiarised from the 1937 film script.

Admiral Byrd Stories - Conclusion

I don't think the evidence supports the notion that Byrd ever travelled to the "inner Earth" - nor did he see any openings! However, stories about this idea continue to circulate, literally "unchecked."

NPIEE - North Pole Inner Earth Expedition

Dr Brooks Agnew

While we are covering one story of a 20th Century Polar Exploration, let's briefly mention the story of a (apparently now defunct) 21st Century Polar Expedition proposal - one which seemed to be specifically focused on the idea of finding an entrance to the Inner Earth. I heard about this in a June 2012 "Coast to Coast" broadcast that featured Dr Brooks Agnew.[77] The expedition's website[78] was launched in March 2011 and from

there we read[79]:

The North Pole Inner Earth Expedition is the Greatest Expedition in the History of the World!...

As a possible motivation for their expedition, they note:

*"Geologists have been aided by Internet linking of seismographic accelerometers to conduct a CAT Scan of the Earth each time there is an earthquake. Of course, like most modern scientists, they mould the data to fit their current paradigm. **The more than 600,000 seismograms have been recently analysed by Dr Michael Wysession[s] and revealed an entire ocean underneath the Atlantic Ocean.**"*

We will be returning to the emphasised statement later. Dr Brooks Agnew director of the NPIEE (pictured above) also suggested that the behaviour of Auroras could be explained by a Hollow Earth:

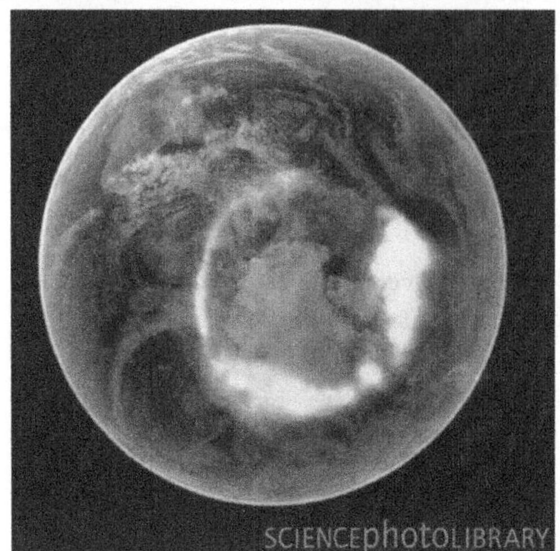

Credit: NASA/SCIENCE PHOTO LIBRARY[80] - Aurora over Antarctica, ultraviolet satellite image. Australia is at upper left.

The Aurora Borealis appears over both poles at the same time. Impossible if it is caused from the solar wind. Very plausible if caused by an ion stream coming from the core of the planet. Besides the precision measurements and samples gathered by state-of-the-art instruments, we will be prepared to test the idea that inside a possible void in the Earth may exist life and an entire ecosystem waiting to be introduced to those of us who live on the surface. In other words, we will be scientifically prepared to catalogue any type of life originating from inside the planet.

The details given on the website state that the expedition would involve a team of up to 35 people[78] over a period of about 2 weeks[81]. According to the website, the planning for the expedition began in 2005, under the auspices of Steven Currey, who sadly passed away the following year. It was after this that Brooks Agnew took over. Though it seemed like a serious attempt at exploration, I have to suggest that it was misguided - as the evidence I have already covered so far (and there would be quite a bit more, I am fairly sure, if you went looking for it) seems more than enough to show that such an expedition would find nothing - other than a whole lot of ice. Perhaps backers of the project also realised it was a waste of time, which ultimately motivated this announcement on 15 Jan 2013, on the short-lived site:

After the sudden and strange disappearance of the Park Avenue, New York film production company that previously had announced its plans to fund the expedition to make a documentary, we have regrouped and decided to run an Indiegogo crowdfunding campaign. If this expedition is to be funded it looks as though it needs to be from the truth seekers of the world, each and every one of us that wishes to know more about our reality! So stand with us and pledge your support for this expedition and go to our Indiegogo campaign...

Their subsequent Indiegogo campaign[82] only raised about $2000 - but it just goes to show, perhaps, that some people will pledge money to almost anything... Do you think I could raise $2000 by encouraging people to watch the 1992 "Pole to Pole" documentary series with Michael Palin...?

6. The Earth Cannot Be Hollow, Can It?

In this chapter we will consider some arguments against the Earth being hollow or having substantial voids below the mantle. In doing so, we will reconsider the diagram/image I used in chapter 1:

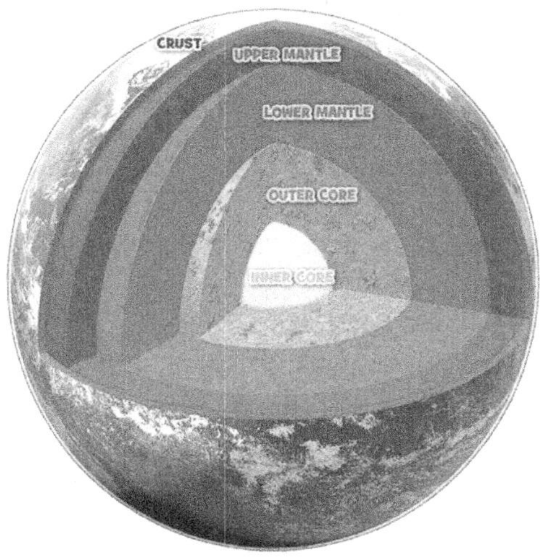

Most people have heard the term "The Earth's Crust" which refers to the outer most layer of our planet - which we live on and can venture into, to some depth. Below this is the hot, partly liquid region called the mantle. When lava comes out of volcanoes, it is coming from the mantle. Geologists report that there is an upper and lower mantle, which go down to a depth of about 3000 km[10]. According to Britannica.com,

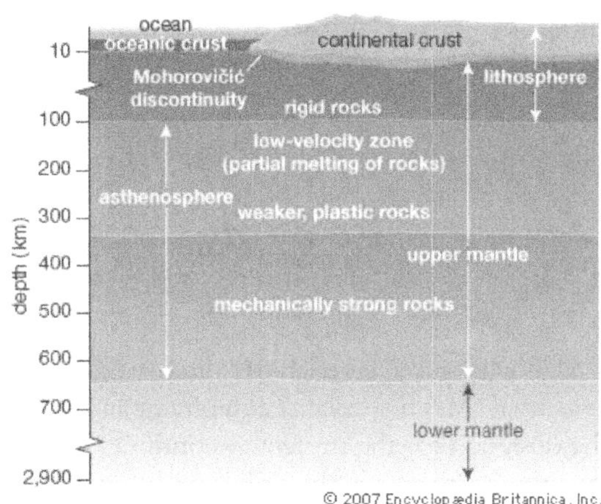

Structure of the Earth - down to the lower Mantle

About one-third of Earth's mass is contained in the core, most of which is liquid iron alloyed with nickel and some lighter, cosmically abundant components (e.g., sulphur, oxygen, and, controversially, even hydrogen). Its liquid nature is revealed by the failure of shear-type seismic waves to penetrate the core.

We will consider the seismic waves again later, but here note that

A small, central part of the core, however, below a depth of about 5,100 km (3,200 miles), is solid iron.

The Britannica article then discusses other aspects of the temperature of parts of the interior - suggesting figures of up to 4,700 °C, but states:

> *Large uncertainties in temperature arise from questions as to which compounds form alloys with iron in the core, and more recent data favour the lower end of the temperature estimates for the inner core. The core's reservoir of heat may contribute as much as one-fifth of all the internal heat that ultimately flows to the surface of Earth.*

It is not clear how the composition of the core has been determined or deduced, but presumably has something to do with the fact that Iron is a dense/heavy element so large amounts of this "sank" to the core. What we can say is that the structure of the core itself has not been deduced from direct observation - for example, through the use of drilling...

The Deepest Hole on Earth

The deepest hole drilled into the Earth's crust up to the time of writing this chapter is the SG-3 Kola borehole. Located in Northern Russia, near the border with Norway, the Kola Superdeep Borehole plumbs down to 40,226 feet or over 7.6 miles (12.3 km) below the Earth's surface.[83] (Some details about this Soviet/Russian project are covered in a "punchy" 4-minute Sci-Show video[84].)

The drilling started on May 24, 1970 with an "Uralmash-4E" and later an "Uralmash-15000" drilling rig was used. Much specialist equipment and several new techniques had to be designed for the project to continue. For example, a method had to be invented so that only the drill head rotated, not the shaft. Drilling continued for about 19 years, but they had reached a depth of 12 km in 1983. At this time, high temperatures were encountered by the drill, which made deeper drilling problematic. Thus, over the next 6 years, they only descended a further 262 meters!

The project itself continued until 1994 and the site remained active until about 2005, when the site was more or less closed down, due to lack of funding - and all the equipment for drilling and research was scrapped. The site has been abandoned since 2008. An article by Alan Bellows posted in 2006[85] contains some interesting details, which are relevant to what we will cover later:

> To the surprise of the researchers, they did not find the expected transition from granite to basalt at 3-6 kilometres beneath the surface. Data had long shown that seismic waves travel significantly faster below that depth, and geologists had believed that this was due to a "basement" of basalt. Instead, the difference was discovered to be a change in the rock brought on by intense heat and pressure, or metamorphic rock. **Even more surprisingly, this deep rock was found to be saturated in water which filled the cracks. Because free water should not be found at those depths, scientists theorize that the water is comprised of hydrogen and oxygen atoms which were squeezed out of the surrounding rocks due to the incredible pressure**. The water was then prevented from rising to the surface because of the layer of impermeable rocks above it.

The article notes that cores drilled from the Kola hole are kept in a repository in Zapolyarniy. These have obviously been studied and the Bellows article reports:

> Another unexpected find was a menagerie of microscopic fossils as deep as 6.7 kilometres below the surface. Twenty-four distinct species of plankton microfossils were found, and they were discovered to have carbon and nitrogen coverings rather than the typical limestone or silica. Despite the harsh environment of heat and pressure, the microscopic remains were remarkably intact.

Hence, we can say that even taking samples just 6 miles/8km or so into the Earth's crust (less than 0.1% of the distance to the centre of the Earth) the structure of the Earth's crust that had been proposed by geologists/planetary scientists did not seem to be correct. So, how reliable can the predictions about what will be down at *2000km* be?

But the Earth Still Can't Be Hollow...

A page on the "Crystal Links"[86] website contains a good summary of issues which are problematic for any "Hollow Earth" theory - what is written below is heavily based on this summary.

Gravity

A strong argument against the idea of the Earth being hollow (or in fact any hollow planet) is the force of gravity. The formation of planets is based on the idea that massive objects tend to clump together because of the force of gravity. It is thought, then, that this clumping would not include large holes - as the force of gravity increases, any holes would be filled in by "inward pressure." The formation of a solid sphere is the best way in which to minimize the gravitational potential energy of a physical object - and so a large hollow would, in theory, never form inside a planet. (We will re-visit this idea later.)

It is also assumed that the sheer mass of the Earth and the associated force of gravity we experience at the surface must rule out the idea of a hollow Earth.

Density

In considering the force of gravity, we should also consider the density of the materials in the Earth. Where estimates have been given regarding the shell thickness of the hollow Earth, they seem to range from 300 km to about 1400 km. We can then calculate the volume of a shell with an outer radius the size of Earth and an inner radius an appropriate size based on the shell thickness. (We will look at this very issue in more detail later). Using this observation or method, the density of the shell material can then be calculated to be 10.5-40.9 (g/cm^3). The density of the crust has been calculated to be is roughly 3g/cm^3[87]. We need to consider the density of Iron is 7.9 and Silicon - which makes up a lot of Earth's mass - is 2.3. The densest of the natural elements are Iridium and Osmium, which are each about 22.6. So, this would mean that for the shell model to be viable, the material in the shell would need to be very dense compared to the materials known to be in the crust, for example. (Elements such as Osmium and Iridium make up a very small part of Earth's mass (which is why they're called "rare-Earth elements"). So, the assumption is, then that for the Earth to have the mass it does, it could not be hollow - as the density of the shell would have to be much higher than would be realistic.

Seismic Information

As we have already considered, the core of the Earth is not directly observable but scientists will say that they have deduced much about the structure from studying "vibrations" - i.e. seismic events - primarily earthquakes - which cause sound or pressure waves to propagate through the planet. Using this method, geologists have confidently stated what the structure of the inner Earth is - consisting of the mantle, outer core, and inner core as we have mentioned earlier. Geologists/seismologists would argue that a hollow Earth would behave entirely differently in terms of seismic observations.

A Re-interpretation of Terrestrial Seismic Waves

Earlier, we mentioned the research of Jan Lamprecht. In an article posted on the "Bibliotecapleyades" website[88], Lamprecht offers an interesting re-analysis of the structure of the inner Earth and he considers how the seismic waves may travel through the Earth differently to how standard models assume. It appears Lamprecht posted this article some time after his book was written, but the original posting has disappeared and it does not seem to have been reposted on his new/updated website. What I have written below is heavily based on his work.

As stated earlier, the inner structure of the Earth has been deduced by scientists through the measurement of seismic waves, which are caused by earthquakes in the mantle (i.e. the region of the Earth below the crust). Seismology is a complex science, so we will only cover the basics here, to help us understand why mainstream science could be wrong about the nature of the Earth's core and folks like Jan Lamprecht could be correct.

There are two main types of seismic waves - "S" and "P" waves[89]. S seems to be used to mean either "surface," "secondary" or "shear" wave in seismology and "P" can mean "primary" or "pressure" wave. An article on Britannica.com explains this in more detail.[90] It contains the following explanatory graphic:

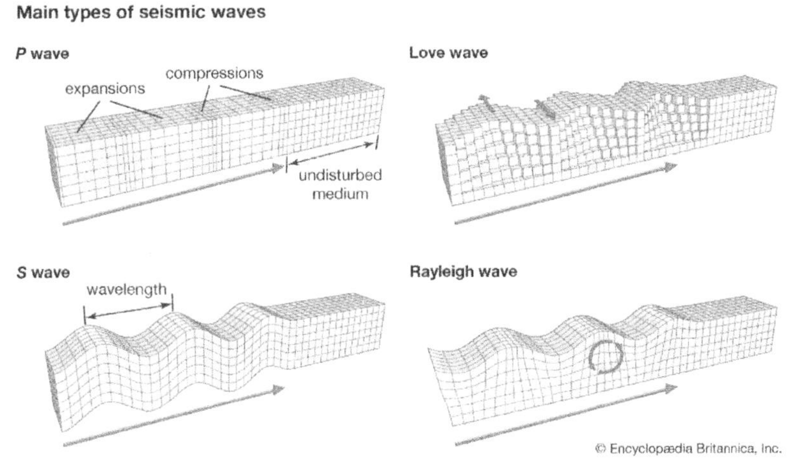

The Britannica article also includes an interesting, short video with an animation showing the propagation of seismic waves through the Earth's interior.[90] The frequency of these waves can be anything up to about 20 cycles per second (Hz), but they are normally much lower. Also, the seismic waves are sometimes called "seismic rays."

Measurement of these waves is made by various seismometers or seismographs which are placed at many points on the Earth's surface. These are monitored by organisations like the USGS (United States Geological Survey) and it is these instruments which can be used in the monitoring and even the predictions of some earthquakes or volcanic eruptions.

Lamprecht references a useful diagram from a 1995 book called "Modern global seismology" by Thorn and Wallace[91] to show what scientists deduce about the Earth's structure from seismic measurements:

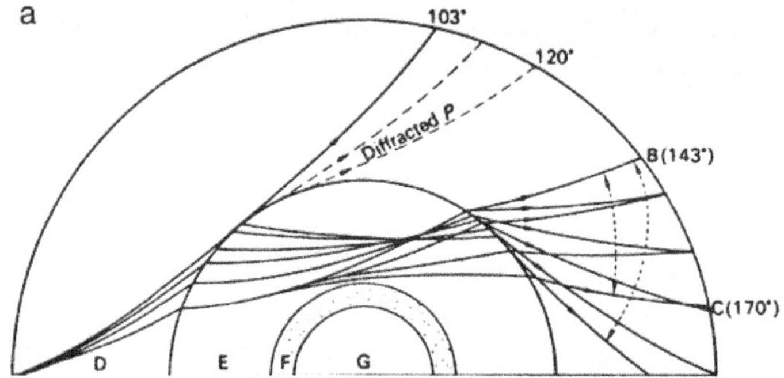

Lines show how pressure waves travel through parts of the Earth where D=Mantle, E=Outer core, G=Inner core.

The diagram is meant to show the situation when an earthquake happens at the bottom left of the diagram - to the left of "D" above. Energy of the quake - in the form of P waves or P rays travels from the earthquake through the Earth. The arrival time, strength and duration of these waves are measured at the Earth's surface by the nearest seismometers (e.g. at the points B and C show on the right-hand side of the diagram). The measurements are collated.

The assumption is that the seismic rays or waves are *refracted* (bent) at the boundary between the different layers of the Earth. At these layer boundaries, the velocity of the P waves increases or decreases. P waves can travel faster than S waves and they can also travel through both solids and liquids. S waves cannot pass through liquids. Under certain conditions, S waves can hit a structural boundary in the Earth and generate P waves.[89]

One phenomenon that seismologists note is the existence of a "shadow zone" on the opposite "side" (opposite hemisphere) of the globe to where the earthquake happened.

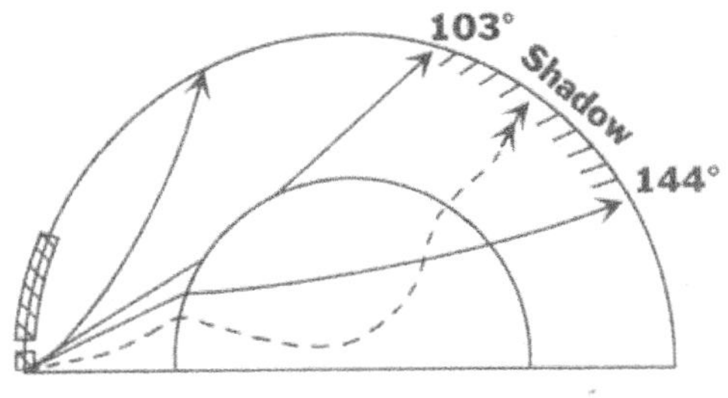

Figure 3.11

Thorne Lay & Terry Wallace; *Modern Global Seismology*, pg 237 (1995)

Again, it is assumed that because the shadow zone can be measured, it proves that part of the Earth's core must be liquid. However, this assumes that the P waves travel along the sorts of paths shown in the diagram and this is why the shadow zone appears. However, Lamprecht argues that the P waves could travel in a different path and never travel through the liquid core, as the diagram above implies. Lamprecht reports that he simulated the path of the P waves and used several different "hollow Earth" structures. He determined that if he used the existing proposed structure of the inner Earth, but with a hollow (or gas-filled) core, he was able to generate the same sort of "shadow zone" that is shown in the figure labelled 3.11 above. He illustrates this idea, thus:

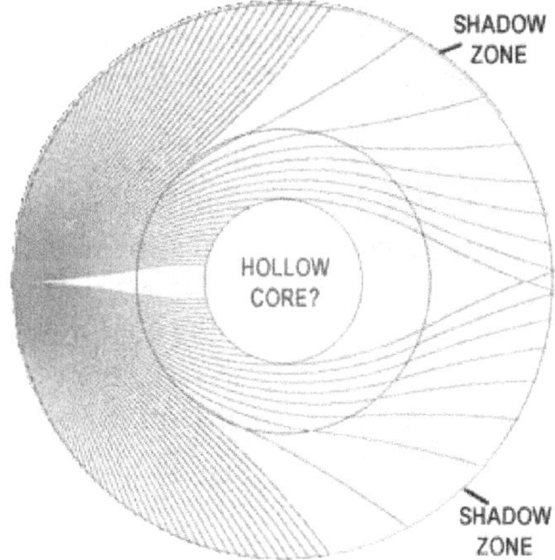

He makes the argument that the P waves could not travel through the hollow core, so instead, some of them travel around it - but in a different path to that which mainstream scientists propose.

He also points out how variations in the velocity of the waves are explained in a way which supports the standard core model - to allow the proposed P wave paths to appear sensible/valid. However, if the path is different, as he suggests, then the velocity changes would imply a different internal structure. It is true to say that we don't really have a precise way of measuring the actual path of the P waves between one hemisphere of the globe and the other.

Earthquake Depths

Another key observation made by Lamprecht relates to the depth at which earthquakes occur. An article on the USGS website on this topic[92] notes the following:

> *Earthquakes can occur anywhere between the Earth's surface and about 700 kilometres below the surface. For scientific purposes, this earthquake depth range of 0 - 700 km is divided into three zones: shallow, intermediate, and deep.*
>
> *Shallow earthquakes are between 0 and 70 km deep; intermediate earthquakes, 70 - 300 km deep; and deep earthquakes, 300 - 700 km deep. In general, the term "deep-focus earthquakes" is applied to earthquakes deeper than 70 km. All earthquakes deeper than 70 km are localized within great slabs of lithosphere that are sinking into the Earth's mantle.*
>
> *The evidence for deep-focus earthquakes was discovered in 1922 by H.H. Turner of Oxford, England. Previously, all earthquakes were considered to have shallow focal depths. The existence of deep-focus earthquakes was confirmed in 1931 from studies of the seismograms of several earthquakes, which in turn led to the construction of travel-time curves for intermediate and deep earthquakes.*

In his article, Lamprecht states:

> *According to scientists, pressure [in the Earth] increases with depth. According to their calculations the pressure is so great that between 70-150 Km down, all rock will begin to flow. Below 150 Km there is no known material which will not flow.*

This implies, then, that if we were to go below 150km into the Earth's surface, we would be dealing with a liquid - not a solid. Presumably, the likelihood of flow occurring in the material increases with depth? How, then, is it possible for some earthquakes to happen in this liquid - or "flowing material"? How can the waves propagate from the "liquid" into solid rock, higher up? Why have no earthquakes been detected below about 700km (430 miles)? Could it be the case that the internal structure of the Earth is not as scientists suggest?

We will revisit this question in a later chapter.

Part 2

The Growing Globe

7. Considering an Expanding Earth

Timescales

In this and later chapters, we will be considering terrestrial events which likely took place millions or billions of years ago. Of course, there are different views on the timescales over which the Earth - and all the life resident on it - formed. I will just briefly mention here the creationist view[93] - which, broadly speaking, takes the descriptions in the biblical book of Genesis as being literally true and therefore the conclusion is that the Earth and its "contents" were created by god about 6000 years ago[94]. Some proponents of the creationist theory offer some interesting and valid arguments. They raise some valid questions about assumptions made by geologists (such as the speed of formation of coal and anomalies in geological strata)[95]. They also make some valid points about problems with the Darwinist theory of evolution. Overall, I think there are too many diverse pieces of evidence (such as many types of radioisotopes, many factors in the fossil record, observation of the speed of a number of geological and terrestrial processes, the motion of the moon and other planets in the solar system) which show that the Earth is, indeed, many millions or billions of years old.

Christian Creationists must also accept that the book of Genesis isn't the only scripture which describes the creation of the Earth and humans - there are, for example, the earlier Babylonian Creation myths (which I wrote about in my previous book "Acknowledged"[96]) and there are other scriptures such as the Indian Rig Veda[97] for example and the less well known Nag Hammadi texts[98].

Having said all this, we must remember, too, that humans only have a lifetime measured in decades - and just ten of those, if we are lucky! It is therefore difficult for us to truly comprehend timescales longer than this.

Plate Tectonics, "Supercontinents" and… An Expanding Earth?

In chapter 1, we covered the generally accepted view of most scientists - that the Earth has been - more or less - constant in size, since its initial formation and cooling, about 4 billion years ago. As regards its landmasses, the consensus view, as we said earlier, was that there was one supercontinent "Pangea" which broke up, and eventually resulted in the arrangement of continents we see today. So let us now consider an alternative explanation for the appearance and arrangements of the continents as we know them today.

As long ago as 1834, the East/West fit of some of the continents, as shown on world maps, had been noticed. For example, the American Transcendentalist poet, philosopher and essayist, Ralph Waldo Emerson[99] made comments about "Continental Shifting" in his lecture "On the Relation of Man to the Globe:"[99]

Some 24 years later, French Geographer and Scientist Antonio Snider-Pellegrini [100] drew the images below (the first known such illustration), in his 'Creation and its mysteries revealed,' (La Création et ses mystères dévoilés, published in Paris in 1858.)

Antonio Snider-Pellegrini's Illustrations - closed and opened Atlantic Ocean, [101]

Dr Alfred Wegener and Continental Drift

It can be argued that in 1889, Roberto Mantovani[102], an Italian Violinist and Scientist, first proposed the idea of "continental drift" - and, indeed, the main idea covered in this section of this book - Earth expansion. It appears that these ideas were later picked up on by Alfred Wegener.

Hence, the idea of a single large continent breaking into smaller landmasses was most famously proposed by Astronomer and Meteorologist Dr Alfred Wegener[103104]. He described his idea of "continental displacement" to the German Geological Society in 1912. He published his research in 1915 in "Die Entstehung der Kontinente und Ozeane (The Origin of Continents and Oceans)." He showed, in the literature of the time, closely related fossil specimens and similar rock strata could be found on widely separated continents - in both the Americas and in Africa. His ideas were ridiculed and his terminology changed to "continental drift" to imply that it was random or "ill-defined." Even in the

1950s, the idea of moving continents was regarded as "ridiculous" by some scientists - perhaps because they could not see or agree on an obvious process which could cause such slow, diverging movement of the continents. However by the 1960s, the Atlantic oceanic ridges had been discovered and these showed that the ocean floor was, indeed, spreading - and so the continents were, indeed, moving apart. Similarly, studies of paleogeology showed that the Earth's magnetic north pole had "moved" - or rather, the continental land masses had moved over where the magnetic pole sat, thus creating traceable magnetic effects in the rocks.

Professor Sam Warren Carey

In the 1950s, Prof S Warren Carey[105], of the University of Tasmania, was one of the people who agreed that Continental Drift (displacement) was real (he was dubbed a "mobilist!". We can read in chapter 8 of Carey's book "Theories of the Earth and universe: a history of dogma in the Earth sciences[106]."

Continental Drift: Mental models bias our thinking, and "continental drift" hobbled Wegener's concept in the English-reading world from the outset. Wegener's word was Verschiebung, which was correctly translated by Skerl as "displacement." "Drift" was substituted by detractors, and as they were the majority, the term gained currency; the theory, saddled with that name, was successfully slanted toward fantasy.

In this same book, on page 96, Carey reports how he first fitted together the continents, in 1933:

Africa and South America as I fitted them at the 1000 fathom iso- bath on an oblique stereographic projection in 1933, before I realized that the Earth is expanding. Although I placed the northwest angle of Brazil as close as I could into the Gulf of Guinea, I could not eliminate the residual gaps tapering in both directions resulting from the expansion of the Earth.

We will discuss the Earth expansion next, but here we will note how, in 1954, he submitted a paper, which included "mobilist/continental drift" research, to the Journal of the Australian Geological Society. The paper was refereed by three professors of geology, one of whom was opposed to the "drift" idea. Hence, their comments were "hostile and disparaging" and Carey said he would never submit another paper to this Journal. However, this paper - entitled "The Orocline Concept in Geotectonics" - was published in 1955 in the Proceedings of the Royal Society of Tasmania. It was widely referenced by "mobilists." Carey also submitted material to some American Geological/Geophysical Publications around the same time. On 12th October, 1971 (by which time the idea of "Continental Drift" - displacement - was widely accepted as being correct), Carey wrote to the American Geophysical Union thus:[107]

Dear Sir,

Twenty years ago at a time when the gross separation of the continents was a heresy to be ridiculed, I prepared a note to answer one of Jeffreys' criticisms of continental drift and submitted it to your predecessor for publication. It was rejected with a note from a referee stating that it was naive and unsuitable for publication. This note contained perhaps the earliest viable exposition of subduction, which is now the accepted dogma, particularly in America. You will also find in my continental drift symposium (published in 1957, although taught by me for many years previously), the first viable exposition of ocean floor growth at the mid-ocean ridges. This is reproduced on page 179 and figure 6 of Search, vol. 1, enclosed. A visiting American geologist recently suggested that I should now send back to you the 1953 note with the suggestion that you might choose to publish it now as a historical document. I therefore enclose a copy of the original manuscript. Although I worked with subduction models for more years than any of the new generation of subducers has yet done, I have since moved on to what I think are more probable models. American thinking has now arrived pretty much at where I was twenty years ago.

Six months later, on 3 May 1972, Carey received the following reply:

Dear Professor Carey

The paper you submitted some time ago to Dr. Spilhaus came to my desk, and I sent it out to a number of members of the Editorial Board, for comments. I asked if we should change our policy and publish a version of a paper we had earlier turned down, however good it may be. The response of the Associate Editors was that we should not change our policy. I regret that the JOURNAL OF GEOPHYSICAL RESEARCH cannot publish your paper.

Sincerely yours, (sgd):

Orson L. Anderson Editor

Carey continued to be "ahead of the game." He had already concluded the Earth was expanding, but he was not the first to suggest this idea or reach such a conclusion. However, it was (and still is) "too much" for mainstream science to accept. Carey's pioneering work on Earth expansion is barely mentioned in his 2002 obituary[108]:

> Observations made during his continental drift studies convinced him that the Earth has expanded. This view has not received general acceptance but does have a vocal group of strong adherents.

I am almost surprised that the word "adherents" was chosen instead of "believers." However, I am not surprised that the word "evidence" does not appear in the paragraph above.

Expanding Earth - History of the Science

In 1888, the Polish civil and mechanical engineer Ivan Osipovich Yarkovsky[109], first proposed that the Earth was expanding - he considered that the Earth was accreting mass, by some "aethereal process"[110] (we will return to this idea later). An Italian scientist, Roberto Mantovani[102] proposed the same idea but, unlike Yarkovsky's theory, this included the observation of the continents separating.

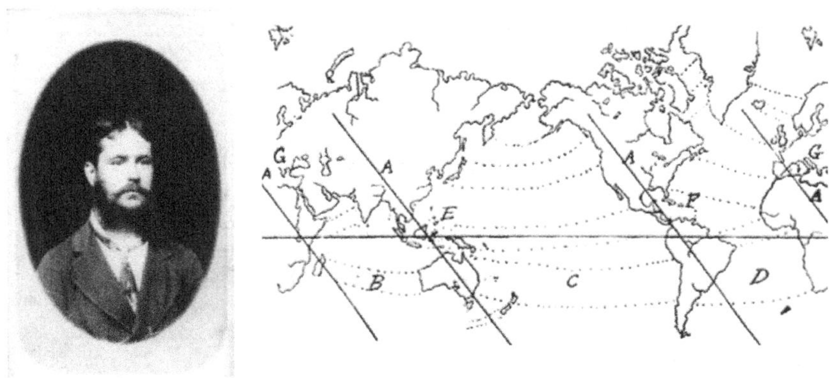

The portrait of Roberto Mantovani beside the map of the opening Pacific in the 1909 paper.[102] The Italian scientist drew the map to show the points on the opposite side of the oceans that were once in contact. The points are joined by dotted lines which were, crucially, were drawn across the *Pacific* Ocean as well as the Atlantic.

It was some years later that Otto Hilgenberg[111] (1896 - 1976), who worked as a Geophysicist in an oil prospecting company, developed a theory of Earth expansion by considering the nature of the gravitational field. In 1933, the theory was published in his classic 55-page work "Vom wachsenden Erdball"[112] ("The Growing Globe"). Hilgenberg included a number of photographs of models of globes he made, showing different points in the Earth's expansion:

Abb. 2. Südamerika, Afrika und Antarktis auf der Schelfkugel.

Abb. 3. Afrika, Arabien, Vorder- und Hinterindien auf der Schelf- kugel.

Hilgenberg's "Growing Globes"

Hilgenberg dedicated his book to Alfred Wegener - who gave him the inspiration. In the book, Hilgenberg developed the original ideas, to include the expansion concept.

A page on engineer Stephen Hurrell's website lists other people involved in Earth Expansion research[113] such as Josef Keindl, Lazio Egyed, Allan Cox, Richard Doell, D Van Hilten, Karl W. Luckert, Pascual Jordan, J. Halm, Lester King, Ludwig Brosske, Kirillow, Cyril Barnett, Kenneth Creer and Ralph Groves. (We will return to Stephen Hurrell's important research in a later chapter.) Hurrell's site also has a useful, comprehensive list of books about Earth Expansion[114].

Here we will note the contributions of Dr Hugh G Owen (formerly of the London Science Museum), who wrote an article titled "The Earth Is Expanding and We Don't Know Why," which was published in the UK's

"New Scientist" magazine on 22nd Nov 1984 (my 20th birthday!)[115] This important article included the diagram below:

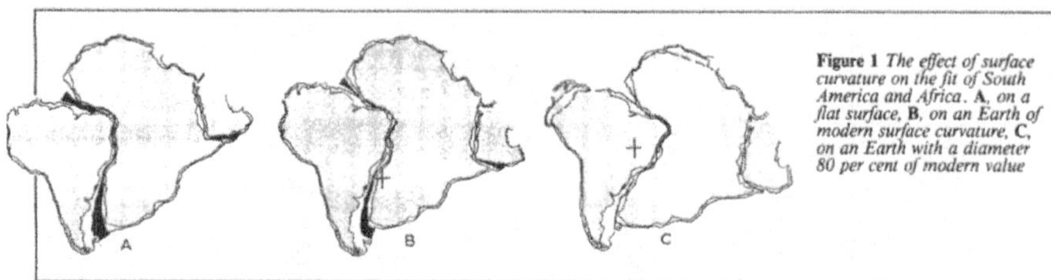

Hugh Owen's version of Carey's "Best Fit" diagram.

Stephen Hurrell established contact with Owen,[116] who then provided Hurrell with electronic copies of detailed continental maps he had developed.[117]

Dr James Maxlow (whom we will talk about later) discusses how engineer and "non-geologist" Klaus Vogel produced some very detailed "terrella" models in 1983[118] - with perhaps the most interesting being a smaller globe surrounded by a transparent one (as shown below). The models show a progressively larger Earth - in a similar manner to the models made by Hilgenberg, 50 years earlier.

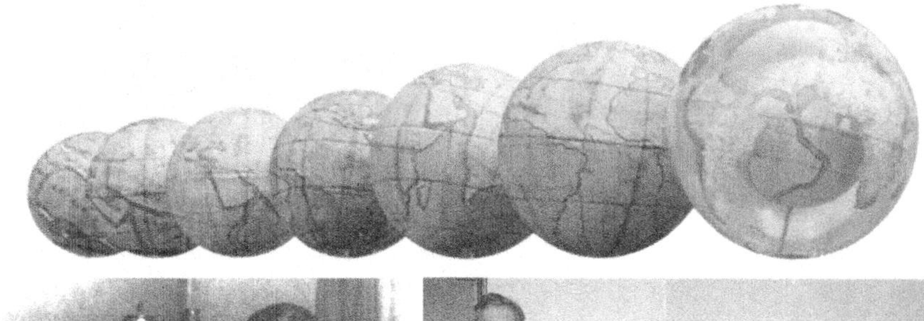

Professor Carey (left) at Werdau, in the German Democratic Republic in January 1979 being shown a smaller Earth within an outer expanded globe produced by Klaus Vogel (right)

Klaus Vogel (left) and James Maxlow (right) displaying and comparing their Expanding Earth models at Klaus' home in East Germany (1997).

More Recent Research - Dr James Maxlow

Perhaps the foremost authority and proponent of Earth expansion evidence and research, as of the writing of this book in Sept 2019, is Dr James Maxlow[119], formerly of Curtin University of Technology, Perth, Western Australia. Maxlow was born in Middlesbrough, England but emigrated to Australia at the age of 4. He started studying engineering at University, but this did not suit him and he switched to a degree in Geology at the Royal Melbourne Institute of Technology, graduating in 1971. Maxlow then worked for over 25 years as a geologist in mining and other industries and positions. He therefore gained very diverse experience. During this work, he noticed an unusual correspondence in the arrangement of widely geographically separated silica and iron ore deposits and sedimentary rocks in the Pilbara region of Australia. This region is a large geological "dome" and Maxlow considered, having done some measurements over an extended period of time, that the shape of this dome could be evidence that the Earth's radius was smaller in the past.

He was eventually able to return to University in the 1990s, where he then completed a master's degree and a Doctorate of Philosophy in 2002. Not surprisingly, he faced considerable opposition in his chosen field of study - Earth expansion. Maxlow's PhD thesis[120], completed in 2001, runs to 451 pages and contains a huge data set. This data set covers areas such as:

- Ancient Climate (Coal deposits and fossils)
- Ancient Magnetic Poles (Paleomagnetics)
- Ancient Geography
- Ancient Biogeography (e.g. the location of Ancient Reefs)
- Modern Space Geodetic Measurements

Maxlow constructed several sets of globes, showing different types of geological maps on their surfaces - this took years of work. From all this data and research, Maxlow had calculated an Earth radius expansion rate of 22mm/year.

Maxlow has written three books on the subject - On the Origin of Continents and Oceans: A Paradigm Shift in Understanding (2014)[121], Terra Non Firma (2015)[122] and Beyond Plate Tectonics (2018)[123]. (I will repeatedly reference the latter work as "BPT.")

Maxlow's Expansion Tectonics

On a separate website - https://www.expansiontectonics.com/ - Maxlow presents a very useful and readable summary of his research, with the following introduction:

In this website, modern geological mapping of the world is used to reassemble continental and seafloor crustal plates on accurate scale models of the Earth with a precision never before attained. For the first time a series of 24 scale models of the ancient Earth is presented, extending across 4,000 million years of Earth's recorded geological history, including one model extrapolated to 5 million years into the future…

My main interest in Expansion Tectonics is driven by the fact that modern geoscientific evidence clearly shows, beyond reasonable doubt, that the concept of an Earth increasing its radius over time far better explains what is empirically observed in geology than conventional plate tectonic theory currently leads us to believe.

Maxlow also offers a free document which contains a lot more detail[124]. He establishes a time-line for the Earth's expansion and, using much more data than was available to all the earlier researchers mentioned, has modelled the Earth's expansion in a much more comprehensive manner. He uses numerous spherical models ("terella"), which are overlaid with geological maps of bedrock. To fully appreciate Maxlow's presentation, an appreciation of stratigraphy - and the agreed ages of the geological strata - would be needed.

The image below comes from section 7.7 of the document above and is accompanied by a description of the formation of the South American continent, as it was split from the other landmasses.

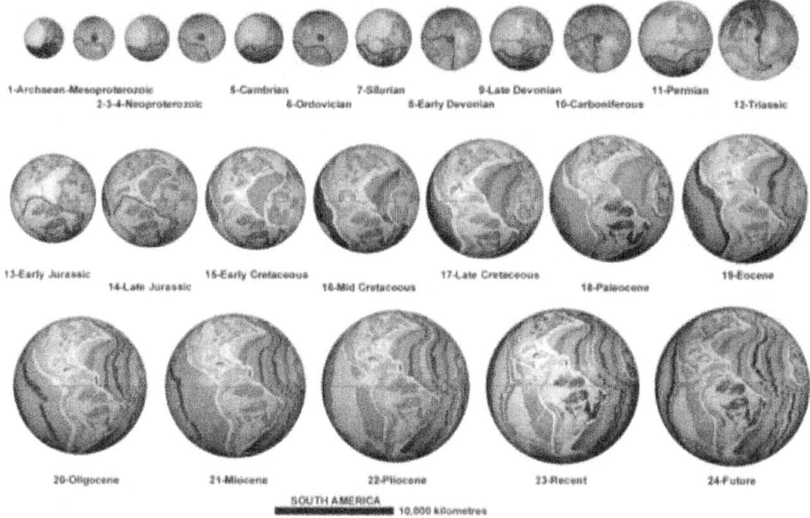

The detail on Maxlow's maps and globes makes the basic picture of the "continental jigsaw" harder to see, but this detail is essential to "prove the case." It is easy to be overwhelmed by the detail, but Dr Maxlow has posted a 2-hour video presentation (from 2005) summarising the key points of his research[125], so this can be viewed to gain an understanding of some of the concepts. We will discuss Maxlow's research further, in a later chapter.

Neal Adams and Modern Graphical Animations

In the early 2000s, Neal Adams[126] achieved some visibility with some compelling computer models/animations[127] of an Expanding Earth - which he originally posted in 2006[128], with some of his own commentary. (I did a podcast/interview with Neal in 2015, where he explained many of his thoughts[129].) Neal Adams was born in Manhattan, USA and is a successful Comic Book Artist - probably most famous for the "Batman" artwork he produced over many years. He has produced highly acclaimed work for most if not all of the major Comic Book publishers. He "cut his own path" in the industry and has gone on to produce work based on characters he has created. He has worked in animation, creating book covers, theatrical costume, stage design and even amusement park ride design.

However, in parallel with this varied work, he has maintained a keen interest in the Sciences, though he does not have any qualifications in a science discipline.

His interest in Earth expansion (not a term he uses) was triggered when he came across the work of Dr Sam Warren Carrey and he realised that the Earth hadn't always been the size it is now.In 2003, he started to post writings and artwork on his website[130] regarding what he came to call "The Growing Earth". Over the next 3 years, he went on to produce some very high-quality digital animations which, illustrated in a

clear and compelling manner, how the Earth has increased in size - over millions of years. On his website, he argues:

Then in a desperate attempt to explain the clear fact that all the continents fit perfectly together geologists say,… that in some magical other unnamed time, and for some unnamed reason, all the continents, once-upon-a-previous-fictional-made-up-time, gathered and connected in the Pacific, again, into one giant island that they named Rodinia. The continents gathered to form Rodinia, in the Pacific, then broke apart and zipped around the planet to gather, and then form Pangea in the Atlantic! You can see why timing is everything in this. How could the two giant islands exist at the same time? What intellectual terror prevents science from the obvious conclusion that Rodinia and Pangea happened at the same time on a smaller Earth, I *cannot explain. This, in the face of facts that the ocean floor in all oceans of the world is the same progressive age and none of it, none, is older than 180 million years old. Apparently it's easier to believe that continents travel around the planet than it is to consider that **the Earth grew**.*

We will discuss Neal Adams' thoughts and ideas later, when we look at how the Earth has expanded - as he has perhaps spoken about this as much as he has spoken about the videos he had created. Neal has a fairly conversational way of talking about the evidence for expansion and how the Earth's structure may be different from what mainstream science claims it is.

For example, in a 4-page document entitled "Gravity And Pressure And Why The Earth Doesn't Have A Molten Iron Core" posted on his website[131] (some years ago), he wrote (on pages 2/3):

All compress iron silicate to a solid shell. Is there such a place exactly like that under the Earth? Yep, it's called Moho. named after the seismologist who discovered it, Andrija Mohorovicic, in 1909. It's the Mohorovicic Discontinuity. It's over 2000 miles down. Sonic recordings of earthquakes around the world note an incredibly abrupt discontinuity of material.

Discontinuity? Yep right down to Moho we have solid material then it changes. There's a small fuzzy area and then something else. Now science says there's a core of liquid iron. I say its gas and plasma. Big difference. I must be wrong.

Ah... not according to science I'm not wrong. Only fuzzy science says I'm wrong, but let me explain. If the Moho discontinuity is as I say, and well, geology says, a true break in material and it is super dense then isn't it really the super thick shell of Earth, a geodesic sphere.

Whether its liquid or plasma below it. Neither are very secure as a base to rest on, are they? It's a solid self-supported shell folks, dense as near solid iron and geometrically supported. I'm asking this question of physicists and engineers. What does science say? At the discontinuity, pressure ends.

Not only does pressure end, gravity is less than one tenth of what it is on the surface. Physicists? In fact, practical engineers, if the core is mostly dense plasma, what is the gravitational attraction? It's nothing, isn't it? Zero! In fact, gravity reverses, doesn't it. It goes outward. Now, let's explore exactly why geologists say liquid iron.

(I've edited the text into longer paragraphs than the original, to aid readability). Adams tends to put his "own spin" on various concepts and ideas - and he does not always reference things very accurately or appropriately. For example, he has, to my knowledge, never referred to the work of Dr James Maxlow or even any of the earlier researchers discussed in this chapter. This meant, for example, that it was quite some time after I came upon Adams' videos that I learned there was, indeed, quite a lot of Earth expansion "research history" in the science/geoscience community!

In the next chapter, we will look at some of the Earth expansion evidence more closely and discuss why some do not easily accept it.

8. Earth Expansion - Expansive Evidence

In this chapter, we will cover the obvious - and less obvious - evidence that the Earth has expanded since its creation. I will quote heavily from Dr James Maxlow's research (because his is the most comprehensive).

Visual Evidence

For me, the most powerful evidence that the Earth has expanded is the same as it was for all the early researchers - that the continental landmasses can be fitted together, almost perfectly, on a smaller globe. This isn't "a theory to be believed" or a "silly belief" - it is simply what is revealed if you look at the existing globe in a different way - and carry out careful observations. Certain measurements can be made, too, which show that the visual appearance is not just an illusion.

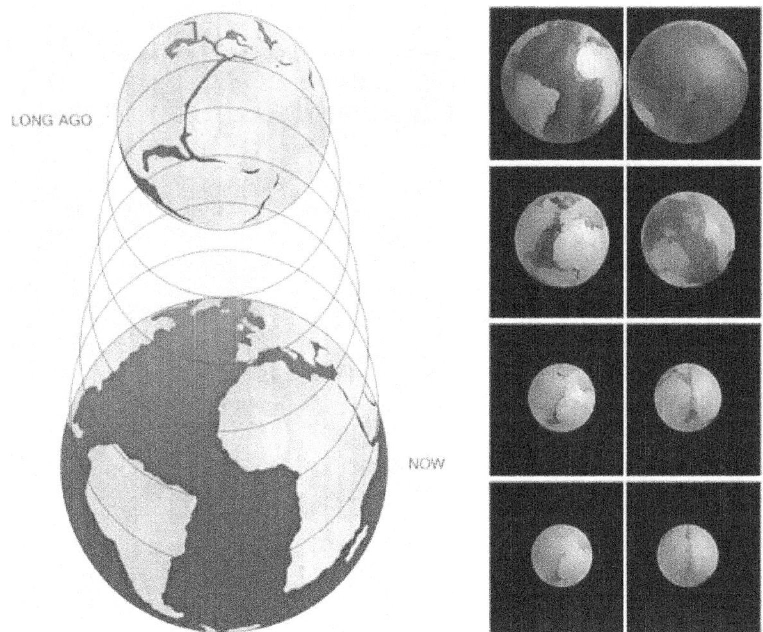

However, visual evidence alone isn't enough - especially when we are dealing with something which is so contrary to what people have been taught - all their lives - about our home planet.

Ocean Floor Spreading

In the 1950s and 1960s, ocean floor mapping was undertaken and, this employed various instruments - including magnetometers - to measure the small local changes in the Earth's magnetic field. It was eventually found that in the mid-Atlantic, the sea floor actually had a striped appearance. The basaltic rock which was formed either side of the mid-Atlantic (volcanically active) fault line was found to have "magnetic striping".[132] A few years later, it was then postulated that the ocean floor was <u>spreading.</u> New material was being pushed out of the volcanically active vents in the seabed, cooling and then forming new ocean floor. As Dr James Maxlow eloquently describes in his "Expansion Tectonics" document:[124]

This seafloor spreading hypothesis was based primarily on the magnetic mapping evidence. It was also supported by several additional lines of evidence available at the time including evidence from age dating and bathymetric surveys. At or near the crest of the mid-ocean-ridges, the seafloor crustal rocks were shown to be very young and these rocks become progressively older when moving away from the ridge crests. The youngest rocks at the ridge crests always have present-day normal magnetic polarity. Moving away from the ridge crests the stripes of rock parallel to the ridges were shown to have alternated in magnetic polarity from normal to reverse to normal and so on. This suggested that the Earth's magnetic field has reversed many times throughout its history. By explaining both the zebra-like magnetic striping and the construction of the mid-ocean-ridge system the seafloor spreading hypothesis quickly gained converts. Furthermore, this seafloor crustal mapping is now universally appreciated to <u>be a natural tape recording of both the history of the reversals in the Earth's magnetic field and opening of each of the oceans.</u>

A profound consequence of this observation of seafloor spreading is that new crust is being continually intruded along the full length of the seafloor spreading ridges. It is interesting to note that this observation was initially - and still is - considered to support the theory of Earth Expansion, where new crust was formed at the mid-ocean-ridges as a consequence of an increase in Earth radius. History now shows that subsequent work has favoured the Plate Tectonic theory, where excess crust generated at the mid-ocean-ridge spreading centres is presumed to eventually disappear along seafloor trenches located along the margins of some continents where subduction of the seafloor crustal rocks is inferred.

We will deal with the "subduction" issue later. For now, we can note that the ocean floor spreading, described above, results in the following:

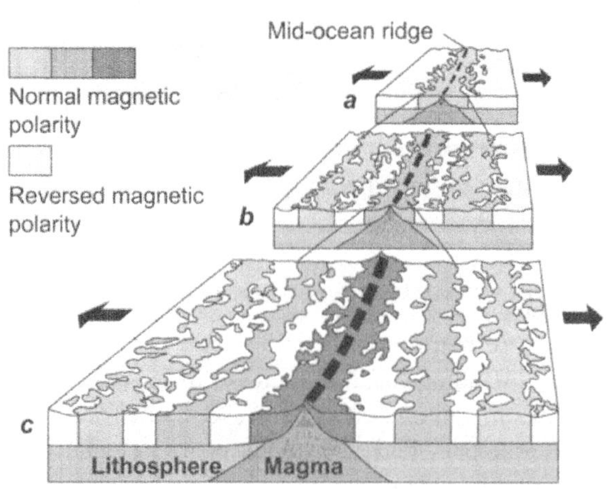

Left: The banding represents repeated changes in the Earth's magnetic field, recorded in the magnetic ores in the basaltic rock crystals. The Lithosphere is another name for "oceanic crust."

Over time, and with more geological data, it was possible to calculate *when* the Earth's magnetic field reversals took place (to within a few thousand years). When this had been done, a map was made and the version below was published in 2008.[133] (You will see various versions of this map online and elsewhere - this one is from the NOAA website).

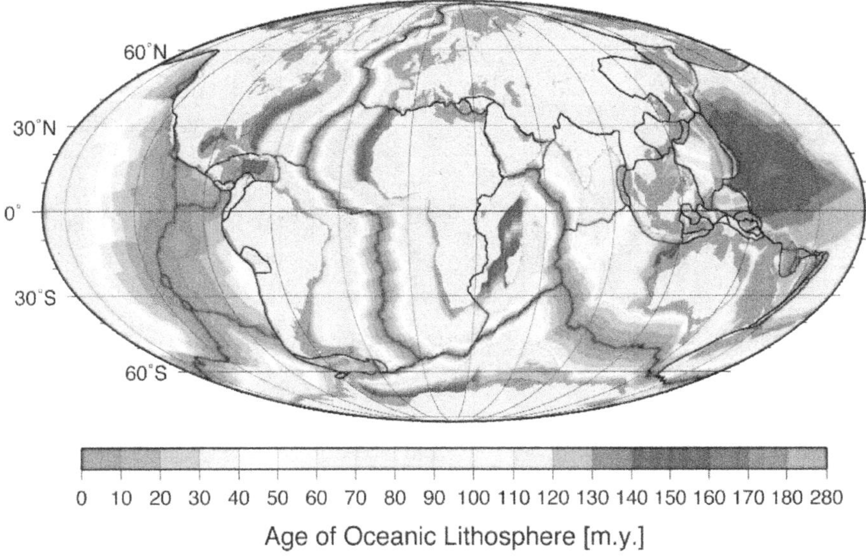

This map reveals that the oldest ocean floor that was found in the deep/wide oceans was about 180 *million* years old. Yet, geologists and scientists had said for years that the age of the Earth was about 4.3 *billion* years old.

Subduction

Before we discuss the Earth expansion evidence in any more detail, it is worth explaining a word we have already seen in this book several times. The term "subduction" is commonly used in Plate Tectonics and the study of the geology of the Earth, it being a primary concept in the paradigm of a fixed-size Earth. We will discuss this more in the next chapter, but for now, we will explain what subduction is.

All geologists now accept that the ocean floor in the mid-Atlantic is spreading - causing "continental drift" (displacement). Hence, if new crust (ocean floor) is being created at one place, crust must be being "destroyed" in another place. Hence, you will see drawings like this on many-a geology website[134] or in many-a text book…

When two oceanic plates collide, the younger of the two plates, because it is less dense, will ride over the edge of the older plate. *[Oceanic plates grow more dense as they cool and move further away from the Mid-Ocean Ridge]. (Image: Keith-Wiess Geological Laboratories; Rice University)*

The older, heavier plate bends and plunges steeply through the athenosphere, and descending into the Earth, it forms a trench that can be as much as 70 miles wide, more than a thousand miles long, and several miles deep. The Marianas Trench, where the enormous Pacific Plate is descending under the leading edge of the Eurasian Plate, is the deepest sea floor in the world. It curves northward from near the island of Guam and its bottom lies close to 36,000 feet below the surface of the Pacific Ocean.

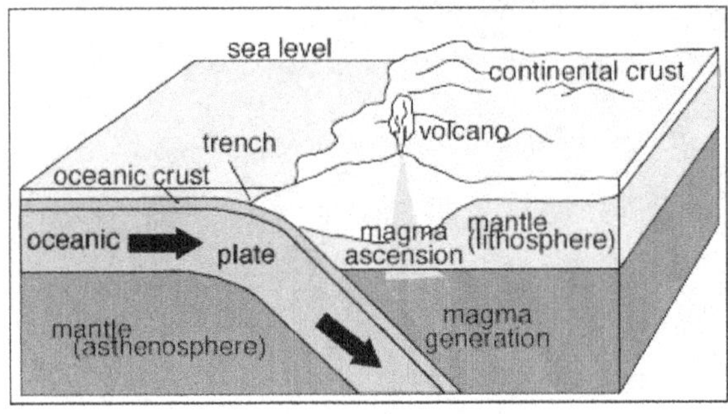

Also, it begs the question as to whether subduction was occurring more than 180 mya, or did it only begin to happen when the Earth's continents began to break up… Again, we will revisit the phenomenon of subduction in a later chapter.

Supercontinents

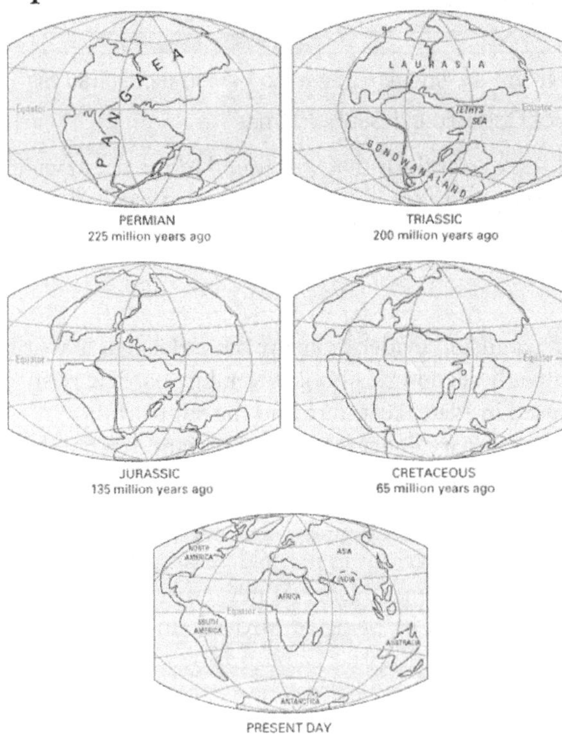

As mentioned above, for the Earth to be constant size, it meant that the oceanic crust must have somehow been "recycled," while the continental crust remained intact - and some of this had to have happened while the "supercontinent" was breaking up. Clearly, we can see that the pattern shown on the NOAA "rainbow" geological seafloor map, shown above, does not match the way in which Pangea was supposed to have broken up (left). It more

obviously suggests… Earth expansion. Revisiting Klaus Vogel's sequence of spheres, we can now understand why his work was important - because he was the first to use modern sea floor mapping data to construct his globes.

Prof Carey had used some of the mapping data in his own research, soon after it became available, but he didn't use it to construct globes, like Vogel did.

For the "fixed radius Earth" plate tectonics model, Maxlow points out several areas that can't be easily explained. For example, the movement of what is now India, as it broke away from the supercontinent, Pangea. Maxlow discusses this in section 6.4 of his "Beyond Plate Tectonics book":[123]

> *India is then inferred to have broken away from Gondwana to drift north as an island continent during subsequent opening of the Indian Ocean. This northward migration of India during the Mesozoic to early-Cenozoic Eras requires* **subduction of some 5,000 lineal kilometres of inferred pre-existing seafloor crust in order to close the ancient Tethys Ocean***. India is then said to have collided with the Asian continent to form the Himalaya Mountains during the Cenozoic Era, leaving behind no trace of this pre-existing ocean crust.*

There is no obvious evidence that this subduction could ever have occurred. Hence, we have an example of where a "fixed radius" Earth model is acceptable to mainstream science even though there is no obvious mechanism for how things came to be - this is the same type of reasoning for which Earth expansion models are rejected! Similarly, in section 6.9, Maxlow observes problems in how the arrangement of land masses in/around the Southern Ocean came to be as it is now:

> *Opening of the Southern Ocean is a paradox on conventional Plate Tectonic reconstructions and very little mention of this ocean is made in the literature. This paradox arises because there are no subduction zones available to absorb the extensive plate motion required to open this ocean or to explain the northward migration of all of the northern continents.*

In "Beyond Plate Tectonics" Maxlow describes the development and motion of the continental landmasses in considerable detail and all those interested in how the Earth came to be as it is today should read his analysis carefully, either in this book, or in the document he has provided as a free download, mentioned above.

Crustal Distortion, Rifting, Collapse after Expansion etc

In his books and writing, Maxlow illustrates the effects on the Earth's crust of an expanding radius. Some of his illustrations are based on the work of earlier researchers such as Dutch geologist D Van Hilten from 1963, who illustrates an "orange peel effect." In the image below, we consider the peel to be sliced or scored, while still attached to the orange (A, left). The peel then represents the Earth's crust. We then imagine that the orange increases in size - e.g. if water is injected into it. Soon, we will end up at "B" and the peel will break apart and leave gaps.

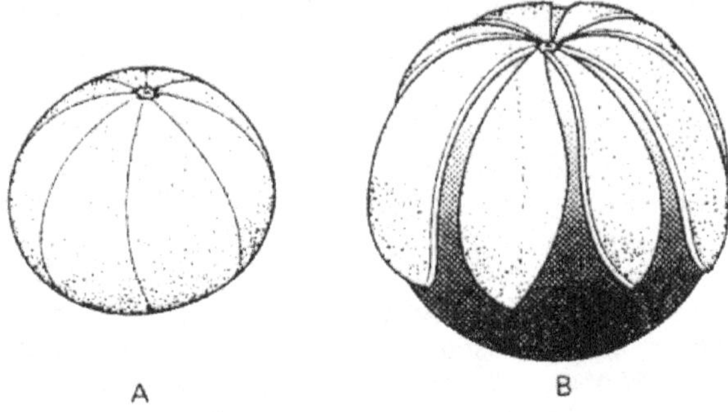

Using an illustration from a paper on "Curvature Expansion" by Australian Geologist M.J. Rickard, Maxlow goes into considerable detail about how the "orange peel effect" would manifest itself on/in the Earth's crust.

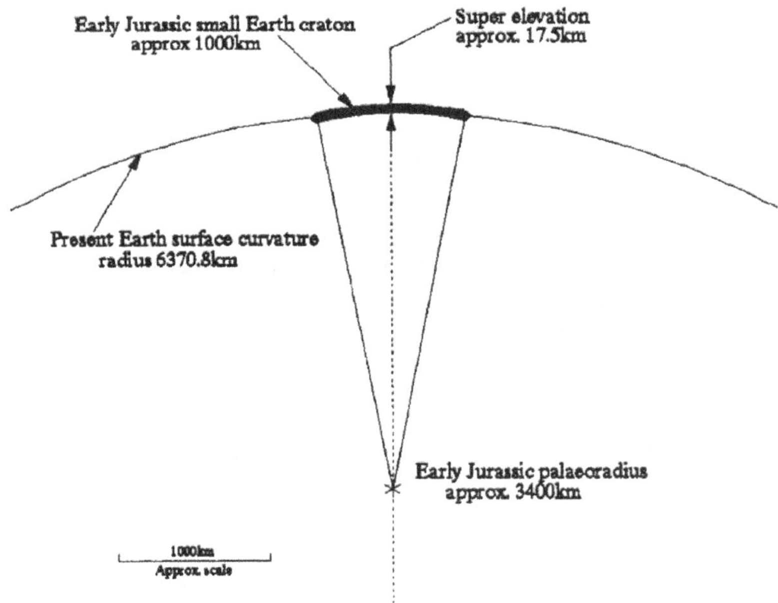

In the diagram above, the emboldened section of crust is called a "craton" - which is a more rigid area of crust. It therefore has a higher curvature - as a result, initially, of smaller Earth radius. At some point, this section of crust is likely to break up and/or collapse under its own weight. This could cause the formation of mountain ranges, basins and other geological or geographic features on the surface of the Earth. Rickard further illustrates this:

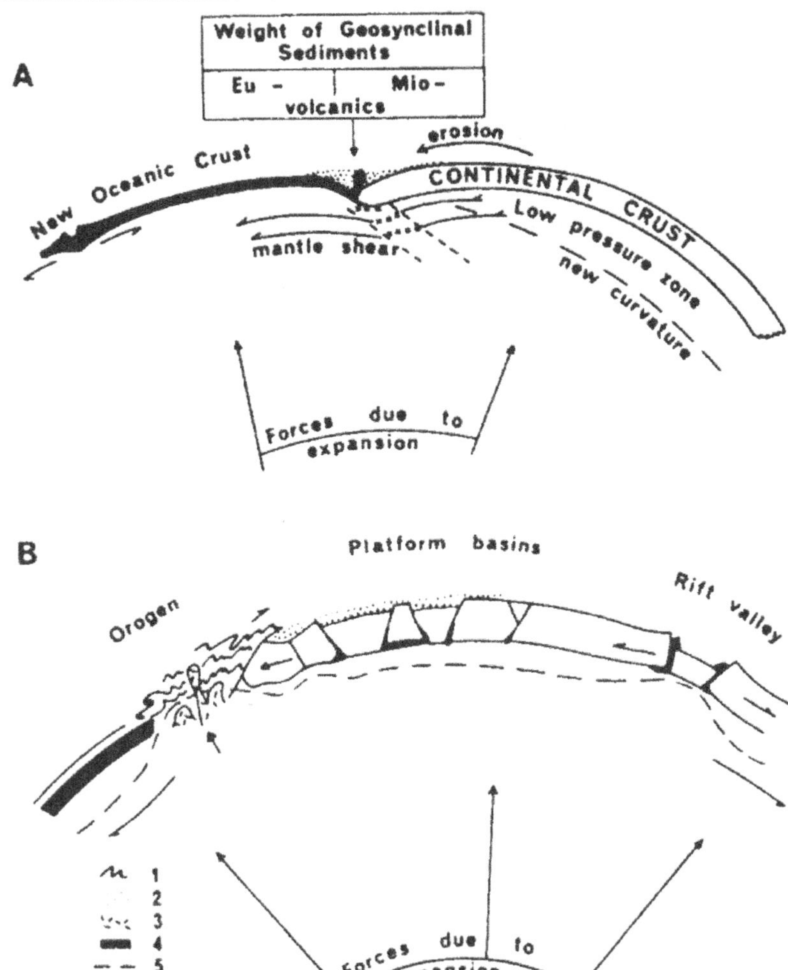

In the diagram above, the change in Earth radius causes stress in the crust which can lead to the formation of "orogens" - where sections of the crust get folded - to form mountain ranges ("orogenesis"). Though this may be an appropriate way to explain the formation of some mountains, a counter argument is that as the sections of crust became raised, they may have been worn down by erosion rather than collapsing to form mountainous features. This is because the rate of vertical movement of such mountains would be relatively slow compared to the speed at which erosion may occur. However, this all depends on many different factors, so no clear answers are available at this point.

Paleomagnetic Pole Data

Again, in "Beyond Plate Tectonics" Maxlow includes a detailed explanation of how geological data indicate that the continental landmasses show traces of where the Earth's magnetic poles were, millions of years ago. He illustrates how a reconstruction of how the poles "migrated" (or rather how the continental plate movement made the poles appear at different points in the various land masses) works much better when the change in Earth radius is taken into consideration, rather than assuming a fixed radius Earth.

Space Geodetic Data

To many people, this term may be quite unfamiliar and it may sound rather technical. However, it simply refers to methods used to measure large areas of the Earth's surface - including shapes. This can now be done using, for example, GPS (global positioning system) devices. As Dr James Maxlow states in chapter 14 of his BPT book:

> *A network of radio telescopes, satellites, and ground-based receiver and transmitter stations from around the world can be used to routinely measure the precise dimensions and continental plate motions of the Earth.*

Methods other than GPS receivers can be used - such as Satellite Laser Ranging (SLR) and Lunar Laser Ranging (LLR). Maxlow argues that in the early 1990s, research by Stefano Robaudo and Christopher G. A. Harrison[135] showed that the Earth radius was "potentially increasing by up to 18 millimetres per year." Unusually, this conclusion was hinted at in a 2002 article in the ultra-rationalist mainstream science journal "Nature," titled "Our planet's waistline is mysteriously increasing."[136]

Maxwell also mentions newer research, thus:

> *Results of Shen et al. in 2011 now show that "...both geodetic and gravimetric observations support the conclusions that the Earth is expanding at a rate of 0.2 millimetres per year in recent decades." This is encouraging but the value is a factor of 100 too low when compared with the current 22 millimetres per year rate of increase in radius based on seafloor mapping data used in this book.*

It seems that Maxlow, unlike other researchers, has tried to use various types of evidence to show what is actually happening, rather than focusing on one or two pieces of evidence and characterising them as "mysterious" or "hard to explain." Maxlow also illustrates how data in the 2011 Shen study is filtered before calculations are made - such that 60% of the raw measurements are "thrown away." Maxlow quotes Shen et al thus:

> *Another concern is that the absolute values of the vertical velocities of some stations are beyond 0.02m/year, and so large vertical movements of such kinds of stations are not related to Earth expansion...Hence, such kinds of stations are not included in our calculations...*

In essence, Shen et al state that geodetic measurements in volcanic regions or where mountain formation is thought to be occurring were not included. I would therefore agree that data showing more significant expansion has been thrown away...

Maxlow also explains that the way space geodetic data is used automatically assumes a fixed radius Earth. i.e. the data are used to measure continental drift/motion, not expansion. Calculations are therefore likely to be used to filter out anything which would show, for example, vertical displacement (height changes). Maxlow correctly argues that more careful consideration or weighting should be given to data showing vertical displacement (i.e. the rising or descending) of the continental crust than most researchers would assign now. In other words, you should look at all the data and modify the model to fit the data - you shouldn't modify (or ignore) the data to fit the model.

Ancient Coral Reefs, Equator and Glaciation

This section summarises another area of study that Maxlow completed. At the start of chapter 17 of his BPT book he points out:

Based on present-day distributions, limestone and coral reefs generally occur within a broad zone located plus and minus 25 degrees of latitude north and south of the equator. The presence of warm water currents may also extend distribution of marine organisms beyond this zone. Within this primary zone, warm ocean waters and currents enable corals and other marine creatures to thrive along the continental shelves of islands and continents. Plotting the distribution of ancient coral reefs on small Earth models will then enable location of the ancient equator...

This therefore allows another set of geological data to be examined - to see if ancient coral reefs and evidence of an ancient equator can be seen across modern day, geographically distant continents. Using data from a 1994 study of Palaeozoic carbonate reefs[137] - i.e. coral reefs that existed between 570 million and 230 million years ago - Maxlow illustrates how reefs can indeed be traced. A study of their differing positions, over time, in the geological record, supports the conclusion that the Earth has expanded, causing separation of the continents.

Maxlow also considers ancient evidence of glaciation, which would typically have occurred in the polar regions. This glaciation produced various types of (eroded rock) deposits, which can be found in parts of the geological record. Maxlow again illustrates that this evidence also supports the conclusion of the continents being joined on a smaller radius Earth, not just in a larger continent or continents on a fixed-radius Earth.

Paleobiogeographic Data

Well, this is another long word - used to describe the distribution of plant and animal fossils in the geological record. Maxlow examines this in chapter 19 of his BPT book. For example, he looks at the distribution of Trilobite, Ammonite, Crinoid and several other types of fossils, using data from PaleoBioDB[138] and plots them on the early Earth models he created. Again, he concludes the data fit better on a smaller radius Earth.

Trilobite   Ammonite   Crinoid

Metallogenic and "Fossil" Fuel Data

Logically enough, Maxlow also analyses the location of various metal ores and deposits - such as those of Lead, Zinc, Gold and Tin and again, these historical mappings seem to fit better on a smaller radius Earth.

It's the same story with Permo-Carboniferous coal, shale oil and shale gas - which formed about 300 million years ago. Plotting the location of these fields on a smaller globe of the appropriate age also seems to be a good fit.

As I wrote in "Climate Change and Global Warming: Exposed,"[139] few people stop to consider, more in the case of crude oil than in the case of coal and similar fuels, whether the "fossil" moniker is appropriate. In simple terms, some of the oil is extracted from depths far below where fossilized remains have ever been found. An interesting page by Col L Fletcher Prouty contains a quote from an August 2002 article/paper, published in the Proceedings of the National Academy of Sciences (US), which had a partial title of "The genesis of hydrocarbons and the origin of petroleum."[140] Dr. Kenney and three Russian co-authors conclude:

> *The Hydrogen-Carbon system does not spontaneously evolve hydrocarbons at pressures less than 30 Kbar, even in the most favourable environment. The H-C system evolves hydrocarbons under pressures found in the mantle of the Earth and at temperatures consistent with that environment.*

In a video interview Prouty contends that in 1892, the Rockefeller family influenced[141] attendees of The Geneva Congress on Organic Nomenclature[142] to conclude that crude oil must be composed of formerly living (fossilised) material - because it consisted mainly of organic compounds. It must therefore be a "fossil fuel" - which could "run out" at any time.

Australian Evidence

In chapter 22 of BPT, Maxlow studies the geological record from parts of Australia and Tasmania, as well as parts of America. Again, he argues that this shows that Tasmania, Australia, New Zealand and the American Continent were once physically connected:

> *Rifting between Australia and the Americas during the Permian Period [between 299 to 251 mya], and rapid opening of the Pacific Ocean during*

the Mesozoic Era, then terminated this established link. During opening of the Pacific Ocean, the Andean and Cordilleran orogenic [mountain] belts fragmented. The resultant mountain belts then remained as part of South America and North America, with the New England fold belt remaining as a small remnant within East Australia, and smaller remnants remain in New Zealand.

Maxlow's Summary

In chapter 11 of his "Beyond Plate Tectonics", Dr James Maxlow writes:

By progressively removing age-dated seafloor volcanic crust from each of the small Earth models in turn it is shown that the global plate fit-together along each of the mid-ocean-ridge plate margins achieves a better than 99% global fit for each post-Triassic model constructed. This unique fit-together is considered to empirically demonstrate that post-Triassic small Earth modelling is indeed a viable process and it is therefore justifiable to consider extending modelling studies back further to the Archaean. This experiment further demonstrates that all remaining continental crusts assemble as a complete Pangaean Earth at approximately 50 percent of the present Earth radius during the late-Permian [approximately 250 mya]

Maxlow also shows, based on several sets of data, that the Earth's radius and its surface area has been increasing exponentially (BPT Section 4.2, 7.3, 13.5)

The Earth is NOT expanding!

Most scientists and academics have never been exposed to the idea that the Earth has expanded since its formation - so they go along with the "fixed size" consensus. One of the main objections is based on the apparent lack of an explanation of *how* the Earth could expand (which we will address later). Another objection is to do with the afore-mentioned subduction.

The Proof of Subduction…?

As we implied earlier, you will hear many geologists and scientists pronouncing that subduction is the process by which plate material is destroyed by going under newer ocean plate. We can read, for example, an article written in 2009 (revised in 2010) by Timothy Casey B.Sc. (Hons.)[110] - a Consulting Geologist. In a section entitled "The Evidence for Subduction: Verified, not Assumed," he writes:

Expanding Earth theory rests heavily on the denial of subduction as a real observed process. Dated sources such as Vine (1987), are often cited as some sort of admission that subduction is the central assumption of Plate Tectonics, when this is not the case. Subduction is verified by plate motion observed at and near convergent plate boundaries and places such as oceanic trenches where seismic equipment is placed and tracked.

Casey then states that measurements near "Wadati-Benioff shear zones confirm as [being] caused by subduction." Casey then goes onto describe a process called "Transport of Cosmogenic Isotopes" where he states that quantities of an isotope of Beryllium, for example, have

...been found to be present in lava erupted in continental and island arc settings is only possible if it is carried down beneath the eruption site by subduction, and subsequently mobilised by partial melting to be incorporated into the source magma.

Casey then argues that research by Tera et. al. (1984), Morris (1991) and others are a sample of studies that therefore provide the "geochemical evidence of subduction."

Further, Casey argues subduction is proved by Plate Motion measured at Oceanic Trenches "by a number of methods including the use of GPS equipment" as described in studies by Holt (1995), Regelous et. al. and others. Unfortunately, he then goes onto state:

It doesn't matter if the intelligent looking man in the suit who says "subduction doesn't occur", is a professor (Carey, 1988), the fact that the definitive plate motions of subduction are measured proves that subduction does occur.

This does not support his line of reasoning! Casey has no explanation for why the oldest ocean floor is aged 180 million years, nor why the continents fit together across the Pacific as well as the Atlantic oceans. As regards the radioactive elements, it is possible that isotopic ratios in materials could be affected by other processes that are occurring in the mantle, so the results he mentions may not necessarily be the result of crust having melted after being subducted. Perhaps I shouldn't mention his characterisation of Prof Carey - which does nothing to bolster his argument for subduction happening in the way proposed.

Dr James Maxlow on Subduction

At several points in his "Beyond Plate Tectonics," Maxlow discusses the problems with the subduction theory. For example, in section 2.1 he writes

Although subduction is now believed by plate tectonists to be the strongest force driving plate motion, it is also acknowledged by many researchers that it cannot be the only force since there are a number of plates, such as the North American Plate, that are moving, yet are nowhere being subducted. The same is true for the enormous European and Asian Plate, and especially the Antarctican Plate.

In section 4.2 of BPT, Maxlow notes that in the past, some scientists have considered a "Partial increase in Earth radius," because it was recognised that there was some crust that, due to the ocean ridge spreading, had to be accounted for.

In section 6.9 of BPT, Maxlow discusses the "Opening of the Southern Ocean," earlier in the Earth's history and notes it:

...is a paradox on conventional Plate Tectonic reconstructions and very little mention of this ocean is made in the literature. This paradox arises because there are no subduction zones available to absorb the extensive plate

Maxlow also notes the appearance of parts of the Geological Map of the World (UNESCO, 1991[143]):

He points out that the symmetrical striping evidence would not appear if there was any significant amount of subduction.

He then argues in BPT Chapter 8 that "subduction-related phenomena are basically related to crustal interaction processes during changing Earth surface curvature." In section 15.5.4 he further states:

*In conventional Plate Tectonic usage the Wadati– Benioff zone is considered to be a deep active seismic zone located within a subduction zone. Motion along this zone produces deep earthquakes, the foci of which may be as deep as 700 kilometres. This same zone and seismic phenomena on an increasing radius Earth is, instead, considered to be related to **obduction** of the continental crust - as distinct from subduction of seafloor crust.*

Using the term obduction, Maxlow suggests the edge of a tectonic plate, consisting of oceanic crust, is thrust *over* the edge of an adjacent plate consisting of continental crust.[144]

Here, I can suggest that, depending on the exact motion of plates during expansion, we may see something that looks like subduction, but actually is not. Due to the timescales involved, it may be difficult to distinguish between plate motions indicating subduction or obduction. Also, we don't know the exact progress of expansion in minute detail and whether there are some inequalities in the internal pressure of expansion. Perhaps the expansion force does not always act equally in all directions (along all radii)? This could perhaps lead to greater underlying expansion in one hemisphere (whilst other forces acting, such as those caused by the Earth's rotation) kept the Earth more spherical. We know that some areas of crust are thicker/stronger than other areas. This would mean some subduction could still occur.

Finally, Giancarlo Scalera writes about Roberto Mantovani's early research into plate motion and how this is understood in "normal" tectonics. Scalera has this to say[102]:

> *The enlarging of huge fractures formed all the oceans. We had to wait [until] the sixties to find the same kind of lines for the Indian and Atlantic oceans in plate tectonics. According to plate tectonics this is not true for the Pacific Ocean, because in this case the plate movement is inverse and the ocean tends towards closing. The 1909 Pacific map was forgotten, and only Mantovani's Pangea representation is reproduced today in some books dealing with the history of science.*

9. "Impossible Fossils"

Dinosaurs and the Expanding Earth

Dinosaurs are a source of fascination, intrigue and wonder for many people and, as a child, I too was captivated when considering how enormous some of them were, based on the fossil evidence.

In this chapter, we will be considering the relative sizes of dinosaurs compared to modern animals. Doing this may throw further light on how the Earth has come to expand. This question has been studied and considered by British Engineer Stephen Hurrell, who has also concluded that the Earth has expanded since it formed. As part of his research, he created his own models of a smaller Earth, but using software rather than hardware! [145]

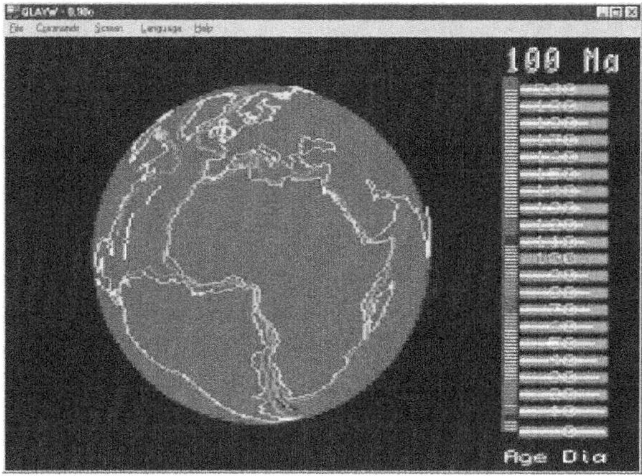

Stephen Hurrell's "smaller Earth" model.

Hurrell created this model by progressively removing the strips of ocean floor as seen in the diagram above. On his website, he writes:[145]

> *Consider how unlikely it is that the entire ancient ocean floor fits together so precisely to form a complete sphere on the ancient Earth. If the missing ancient ocean floor had been generated by any other process than an Expanding Earth it would be improbable that the areas were the exact shapes required to reconstruct a smaller Earth. Surely it would be more likely that irregular shapes that didn't fit together would exist. It is similar to arguing that a jigsaw puzzle fits together by chance rather than for any logical reason.*

He also includes a very useful pair of "GIF" animations - these show something very similar to the "more impressive" animations that were produced by Neal Adams.

In 1994 Hurrell published a book called "Dinosaurs and the Expanding Earth." A third edition of this book was published in 2011[146]. Hurrell also presented a paper at a 2011 conference on Earth expansion.[147] In the book and the paper, Hurrell presents evidence and calculations illustrating that the most likely explanation for the enormous size of dinosaurs (and other life) is a

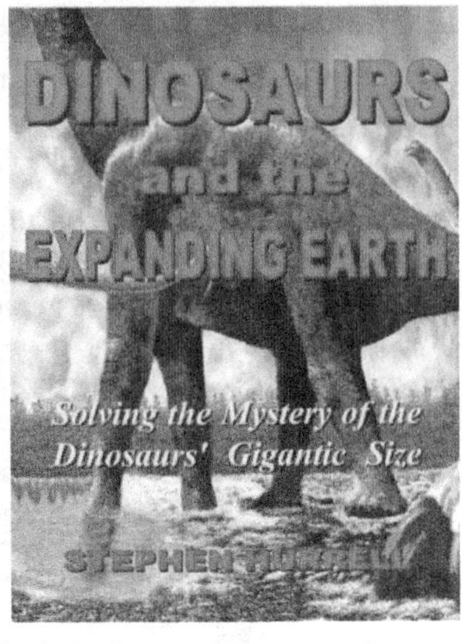

reduced force of gravity at the Earth's surface. Such a concept is, of course, a scientific heresy - because there is no known way that this could have arisen. According to conventional thinking, the Earth has remained fixed in radius and of relatively constant mass - hence the force of gravity experienced at the surface does not change appreciably!

In the diagram below[148], Hurrell illustrates how weight increases with volume and so an animal of double a given length would have eight times the body volume (and therefore, potentially, eight times the weight). This means that the stress on the surface area of any load-bearing bones would be twice as great.

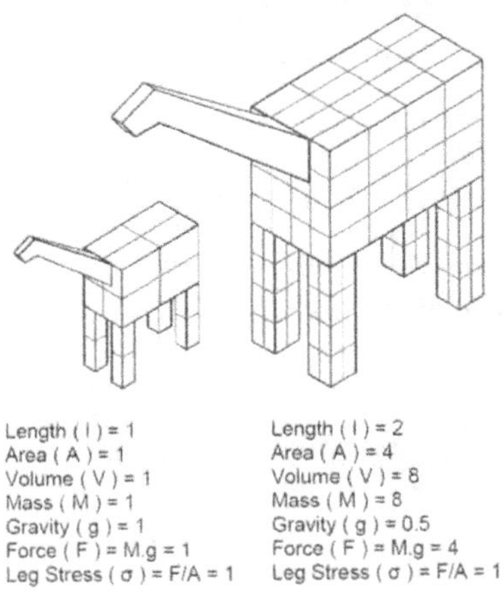

That is to say that weight of an animal would increase *cubically* while the load-bearing ability of the bones increases by *the square of the length*. These factors affect the maximum size for all forms of life, due to the mechanical stresses experienced in bones or other rigid biological elements.

In the diagram on the left, Hurrell is showing how the force on a given surface area of bone in the two animals could only remain the same if the force of gravity acting on the larger animal was *half* the force acting on the smaller one, to keep things "possible."

In his book, Hurrell describes several examples of different dinosaurs such as Diplodocus and Brachiosaurus and how their size and weight can be considered anomalous, if we assume that the force of gravity has remained the same since the time they roamed the Earth. For example, Hurrell points out that from the fossilised bones of the Diplodocus, it can be calculated that the neck bones would not have been sufficiently strong to hold up the head of the animal, if it had lived on the Earth today.

A stark comparison of the size of some of the dinosaurs and some of the largest animals on the Earth today is shown below.

I shouldn't need to state the obvious about the blue whale being an ocean-dwelling creature!

Gigantoraptor erlianensis - A Case Study

In 2019, Stephen Hurrell posted a well-researched paper on his website discussing weight calculations for Gigantoraptor erlianensis - a large, bird-like dinosaur[149] (but not necessarily one capable of flight). Hurrell's study is based on a fossilised creature unearthed in Mongolia in 2010 by a Chinese Palaeontologist named Xing Xu. The find is thought to date from about 80 mya.

In his paper, Hurrell notes some of the problems that arise with calculating the weight of dinosaurs when they were alive:

> *Unfortunately there is still a great deal of confusion between weight and mass and this has resulted in some palaeontologists trying to produce low mass estimates to conform to weight. Paul (1988, p130[150]) for example explains how he used weight calculated from bone dimensions "to expose implausibly high mass estimates ... so a higher mass estimate should be examined critically." All this general confusion between weight and mass has undoubtedly reduced many mass estimates to unreasonably low values.*

Hurrell then considers the more general concept of tissue density in animals and also considers the volume of an animal's lungs which affect the average density of an animal when its overall volume is taken into account:

> *Similar reasoning implies that the tissue density excluding the lungs is 1.03 tonnes per cubic metre, not the 1 tonne per cubic metre often assumed for these calculations. Many studies also assume that there were additional isolated air-sacs within dinosaur bodies to reduce their mass. However, the buoyancy effect of the lungs means that living animals can float in water because they are slightly less dense while a drowned animal sinks in water once the lungs are full. Since dinosaur fossils are often recovered from the bottom of ancient rivers or lakes it would indicate that their tissue density was similar to today's life when they drowned. It would therefore seem unlikely that dinosaurs contained any isolated air-sacs that reduced their mass by a substantial amount.*

Hurrell sensibly contends that body tissue density of dinosaurs was about the same as that of modern-day animals. Hurrell's paper describes apparent attempts in paleontological literature to "fudge the data" (my characterisation, not Hurrell's!) - to ensure that dinosaur weights present none of the problems that I have mentioned in this chapter:

> *Since the bone results were published in 1985 the mass of dinosaurs based on volume methods have been reduced to try to agree with these super-lightweight estimates for dinosaurs. Since the two methods give very different results some palaeontologists, as noted previously for Hutchinson et al (2007), advised abandoning the use of the formula based on leg bones entirely, since they cannot get dinosaurs' mass small enough to agree with the bone weight calculations. These types of criticisms encouraged Campione et al (2012) to slightly modify the original Anderson et al (1985) formula to produce increased weight estimates for larger dinosaurs more in line with the volume mass estimates.*

In his paper Hurrell, using calculations based on the "vital statistics" of the Gigantoraptor erlianensis, suggests that the force of gravity that was extant 80 mya was about 60% of what it is now.

In his 2011 book and his additional papers, Hurrell correctly (in my view) concludes that the only way that large dinosaurs could have existed is by being subject to a lower force of gravity. We will discuss how this fits in with the Earth expansion evidence in a later chapter.

"Under Pressure"

Before we consider the problem of the aerial dinosaurs, let us pause to consider just one of many issues raised on David Esker's interesting "Dinosaur Theory" website[151]. He notes the problem of blood pressure of the Brachiosaurus:

> *Many researchers have questioned how it would be possible for a Brachiosaurus to supply blood to its head. Several unlikely hypotheses have been suggested. Some palaeontologists have suggested that Brachiosaurus had a massive heart to produce the needed pressure to lift the blood.*

> *Another proposal is that the Brachiosaurus evolved a series of several evenly spaced hearts in the neck as a pumping system that would get the job done. More recently a popular idea is that the Brachiosaurus never lifted its head up but instead just moved it back and forth horizontally.*

Esker includes some calculations relating to blood flow and blood pressure. He also eloquently considers the blood circulation in a giraffe, thus:

> *Yet the giraffe's greatest cardiovascular problem is having a strong enough heart to lift blood up to its brain. To produce the necessary blood pressure the giraffe's heart is a huge muscle with walls up to three inches (eight cm) thick and weighing 25 pounds (11 kg). But even more impressive is that the giraffe's resting heart rate is 65 beats per minute. This is about twice what is expected for an animal of its weight. The giraffe's massive 'revved up' heart produces the 300 / 180 mm Hg blood pressure needed for the blood to reach the giraffe's head. Giraffes have a relatively short lifespan of only 20 years and are prone to heart attacks as a consequence of their cardiovascular adaptations.*

Of course, these same issues would be greatly magnified for the larger dinosaurs.

Esker, like Hurrell, considers the bone and muscle strength of dinosaurs and includes a table of figures relating to the stresses/forces present in the bones of existing mammals. Not surprisingly, the smallest stress is experienced in the bones of a meadow mouse and the largest in an elephant.[151] Esker also considers several other areas where the size of dinosaurs seems to "break the laws of physics." Nowhere is this perhaps more obvious and problematic for "conventional thinkers" than with the flying dinosaurs.

The Flying Lizards

Whilst some scientists and researchers have argued that the larger dinosaurs must've spent a considerable time wading in water - to support their own weight, due to the problems described earlier in this chapter, this argument or suggestion cannot be applied to those dinosaurs that became airborne.

The fossil record clearly shows some dinosaurs were the giant ancestors of modern birds, but how did they have the "power to weight ratio" in their anatomies to keep them aloft?

David Esker writes about one of the lesser-known large flying dinosaurs thus[152]:

> *Quetzalcoatlus[153] - Unlike the Argentavis there are no living relatives of the Quetzalcoatlus and this makes it difficult to estimate the Quetzalcoatlus' mass. The author produced values that fell in a range between 500 kg to two tons; thus arriving at a rough estimate of 700 kg. With a 12 m wingspan and a chest cavity larger than that of a horse there is no getting around the fact that this was a huge animal.*

> *However some palaeontologists tell a different story in estimating the Quetzalcoatlus mass to be between 90 and 250 kg. Thus they are claiming that the Quetzalcoatlus had a body density that was about seven times less*

On the page referenced above, Esker also includes some figures and calculations to more clearly illustrate the problem of how these creatures could have left the ground, considering how heavy they would have been.

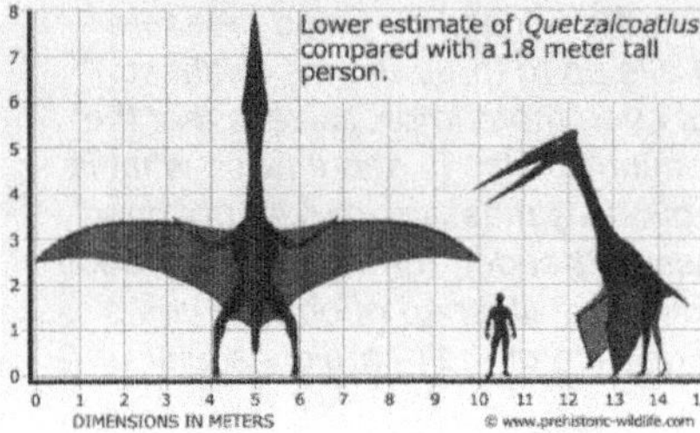

Compare the weight of even 90kg to the weight of one of the largest modern-day flying birds - the Kori Bustard, which can be up to 19kg[154]. The Quetzalcoatlus[155] weighed, at a minimum, over four times as much! Using the upper estimate of the pterosaur's weight would mean the Bustard is *ten times* lighter than its prehistoric predecessor. We have similar problems with the morphology of creatures like Argentavis[156], although its wingspan was only about 7m.

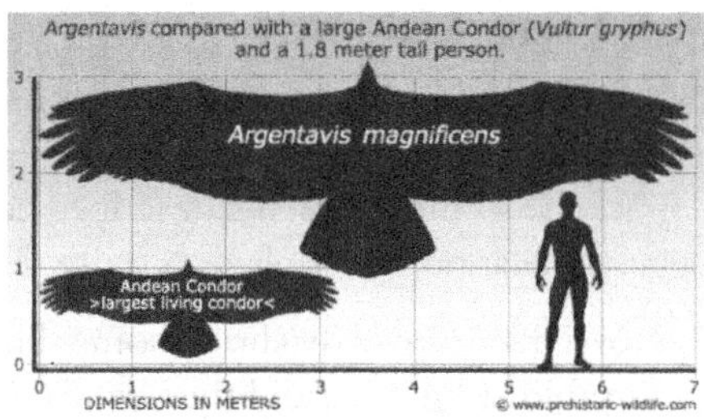

Larger than (Current) Life

Life-size replica of Arthropleura, largest land-dwelling arthropod which ever existed. Image credit: James St. John.

One might argue that there was something "special" about dinosaurs which allowed them to grow to enormous size. Perhaps there was some unknown property of their bones, tissue or bodies which made them large but not very dense/heavy? Perhaps palaeontologists would pursue this line of thinking if it weren't for the fact of the existence of other types of large prehistoric creatures. On a site called "Prehistoric Park Wiki,"[157] we can find the following text:

Arthropleura was a relative of centipedes and millipedes. It grew to over 2 metres in length. Arthropleura is the biggest land arthropod known to man… Arthropleura lived in the carboniferous forests, approximately 340-280 million years ago.

But why didn't the exoskeleton of such creatures collapse under its own weight…? How did the large insects remain mobile?

Another fossilised insect has been discovered in several places - and it has a direct and quite familiar analogue in modern day fauna. Meganeura is described and shown on "Prehistoric Park Wiki:[158]"

Meganeura was an eagle-sized dragonfly, with a wingspan of up to 75 cm (2.5 ft.) that died out by the end of the Carboniferous time period. Like the modern dragonflies, it was carnivorous.

The site shows the image (left) of someone called "Nigel" holding a scale-model of the creature.

Just imagine coming across a similarly sized insect when you were picnicking by the river! The "Prehistoric Park" site also notes:

Upper Carboniferous air is 35% oxygen, not 20% as now, and that is why the insects are so big.

This is also stated in an article entitled "The History of Air" on the "Smithsonian Magazine" website[159]. The article discusses how a combination of giant plants and inefficient bacteria (which would normally consume oxygen) led to higher oxygen levels back then.

Whilst higher oxygen levels may explain the reason why insect bodies could still breathe (respire) effectively when they were several times the size of modern day insects[160], it does not explain how Meganeura was able to fly…

Size DOES Matter...

This chapter has attempted to illustrate that the force of gravity experienced at the surface of the Earth must have been smaller, during the time of the dinosaurs. This is the conclusion that Stephen Hurrell has reached and I agree with him. However, we must also include comments by David Esker who has also done important calculations concerning the weight and "flying ability" of the dinosaurs that we also mentioned in this chapter. David Esker has made the very unusual suggestion that in the time of dinosaurs, the atmosphere of the Earth was *much* denser/thicker in order to support the weight of the animals[161]. He states:

> It may be difficult for some people to imagine how the Earth could have had such a dense atmosphere. But nevertheless, the wonders of our reality often exceed the limitations of many people's imagination. Esker's Thick Atmosphere Theory violates no property of science. It is the correct solution.

Whilst he provides some calculations to support this idea, he does not explain how animals might have been able to breathe in such a dense atmosphere. Further, it occurs to me that within the brachiosaurus, the forces on the blood circulating in the body would not be changed by a denser atmosphere outside the animal. Arguably, if the atmosphere was denser, there may even be greater external pressure on some blood vessels, which would further inhibit blood circulation. Hence, it seems that Esker's confident assertion regarding the "correctness" of the "thicker atmosphere theory" is unfounded - as it ignores certain parts of the evidence he himself cites to illustrate the problems with dinosaur physiology shown in various fossils. Also, it is not clear whether he has considered the aerodynamic effects of a thicker atmosphere on the flying dinosaurs, which he also discusses - then seemingly ignores in drawing his bizarre conclusion.

So now, having shown compelling evidence that the Earth has not remained fixed in size and that some prehistoric animals species, including dinosaurs, became "impossibly large," we can look further at the reason Earth's radius - and the force of gravity experienced at the surface - have increased.

10. Explaining Earth Expansion

If you read sources like Wikipedia, and just about all standard academic works examining the plate tectonics theory in any detail, where Earth expansion is ever mentioned, it is likely that it will be rejected as a correct explanation of the facts, simply because mainstream science cannot "conceive of" or imagine a mechanism which could generate such expansion.

In his writings, Maxlow repeatedly quotes an associate - Polish Geologist Stefan Cwojdzinski[162] - who wrote to him in 2005:

> *The insinuation that we still do not know a physical process responsible for an accelerated expansion of the Earth is not a scientific counterargument... It is not a task of the geologist to explain problems beyond their discipline. Their task is to see and correctly explain all geological facts."*

It seems that with all bodies of evidence which challenge an established consensus view or theory, believers of the mainstream theory or opinion immediately jump to the "how?" question before fully considering the "what" question. That is, there is a tendency to reject the evidence when there is no obvious way to see "how the evidence could have got here." At this point, we can quote Leo Tolstoy[163]:

> *I know that most men not only those considered clever, but even those who are very clever, and capable of understanding most difficult scientific, mathematical, or philosophic problems can very seldom discern even the simplest and most obvious truth if it be such as to oblige them to admit the falsity of conclusions they have formed, perhaps with much difficulty conclusions of which they are proud, which they have taught to others, and on which they have built their lives.*

In this chapter, we will cover some of the theories which have been proposed by several people, to explain how the Earth has expanded or "grown." Following this, in the next chapter, we will explore in depth (no pun intended) one particular theory which I think has the most evidence to support it being the correct explanation for how and why the Earth has expanded.

Some Earth expansion explanations that have been put forward by Maxlow and others are:

1. A pulsating Earth, where cyclic expansion of the Earth is said to have opened the oceans and contractions have caused mountains to form
2. Meteoric and asteroid accretion - expansion is caused by an accumulation of extraterrestrial debris over time.
3. Constant Earth mass, with phase changes of an originally super-dense core.
4. Change (reduction) in value of the universal gravitation constant G.
5. A "cosmological cause" resulting in an increase in the mass of the Earth.
6. Marvin J Herndon has proposed a "whole Earth decompression dynamics" theory.

Also on James Maxlow's website[164], he gives a summary of the problems with the first 5 of these theories, noting how they were considered by the foremost authority on Earth expansion, Prof Sam Warren Carey (whom we discussed in an earlier chapter).

Pulsating Earth

This is where cyclic expansion (and contraction) of the Earth is suggested to have opened up the large oceans whereas the periods of contractions caused the formation of mountains. However, as Carey observed (Maxlow agrees), there is no evidence of this "pulsation" in any of the modern ocean floor mapping data. Similarly, it cannot explain the observed exponential expansion.

Meteoric and Asteroid Accretion

This is the theory favoured by some researchers such as Stephen Hurrell. Perhaps this is because of what Maxlow writes in chapter 1 of BPT:

> In 2002, Koziar showed that even though Earth mass and universal gravity are assumed to be constant for space geodetic purposes, the incremental change in Earth mass can be deduced from space geodetic observational data. The precise measurement of G.M [the product of the gravitational constant and the Earth's mass] began in the late-1970s and in his review Koziar took into consideration measurements that continued into the 1990s. This space geodetic data were shown to consistently record a slow increase in Earth mass of the order of 3×10^{19} grams/year.

This increase in mass can almost certainly be attributed to meteoric dust.

In chapter 5 of his book, Hurrell addresses the reason for the increasing force of gravity at the Earth's surface and states that he believes the mass has increased by "eight times in the last few hundred million years." In his conclusions in chapter 7, he writes:

> ...we looked at how the Kant Laplace Nebular hypothesis for the creation of the Earth could be modified to account for the expansion of the Earth over geological time. The formation of the Earth was not an initial rapid bombardment of cosmic material but the much slower process of an ongoing cosmic accretion from asteroids, comets, meteorites and cosmic dust. This process has lasted for the geological history of the Earth. Today we only see some cosmic showers, but over geological time there have been many cosmic storms when the Earth has been bombarded by vast amounts of cosmic material. Most of this cosmic material has been transported into the interior of the Earth by the action of weathering, material transport and subduction.

But if this meteoric/cosmic dust accretion was constant, would we have more consistency between, say, the appearances of bodies like Mars, the Moon, mercury? Though the Earth has an atmosphere and oceans, wouldn't the geology of the Earth be much more uniform if the mass accretion process had added over half of the Earth's current mass? Wouldn't we find it much harder to find differences in the elements found on the Earth to those found in meteorites. (i.e. meteoric rock is distinguished from Earth rock by its elemental composition). It seems clear from observation that if there is/was a steady stream of meteorites and cosmic dust, like Hurrell proposes, it would gradually obscure any sort of zoned-geological differentiation - such as is found in the global geological map, seen earlier. If Hurrell is proposing that about seven-eighths or over 80% of the Earth's current mass has accreted from the solar system/cosmos, then one would expect to see very different geology overall. Similarly, it is a lot harder to explain the exponential increase in Earth radius since it started to expand - as this would imply a similar exponential increase in meteoric/dust accretion - for which there appears to be no evidence. Similarly, Hurrell suggests the meteoric dust is carried into the Earth's mantle by subduction, but if we assume this happens on the scale needed to move the material there, we are back to the same problem we have with the fixed-radius Earth - we need subduction to account for the disappearing material!

Whilst this theory does tie in with the various extinction events that have been recorded in the geology, Maxlow states that Carey rejected "meteoric matter accretion" as the primary cause of Earth expansion, since expansion should then decrease exponentially with time (as the bombardment by meteors seems to have decreased, not increased).

As Maxlow also states, "added mass" in the way Hurrell suggests does not explain ocean floor spreading, or the distribution of oceanic crust or the covering with sediments. Maxlow further considers the gravity issue in chapter 7 of his BPT book. Though he illustrates problems with both the fixed Earth mass and increasing Earth mass scenarios, Maxlow tends to "come down" on the side of the increasing mass scenario, due to the observations of the *smaller amounts of* mass accreted from meteoric and other cosmic dust. Maxlow also suggests the solar wind could play a role in increasing the Earth's mass, which we will cover in the next section.

Constant Earth Mass

Mainstream science sticks with this assumption and ignores the expansion evidence that has been laid out by all the researchers (and others) mentioned in this book.

In the next chapter, we will explore how the Earth can expand without changing in mass all that much. For those who accept Earth expansion, but not mass increase, they have to resort to some more exotic explanation of the change in size without a change in mass - which involves "phase changes" in the matter of an originally super-dense core.

Maxlow states that this sort of explanation was also rejected by Carey as the main cause of Earth expansion because it would mean surface gravity (and density) was too strong throughout the Precambrian to Late Palaeozoic Eras.

(Maxlow reports that a large Precambrian surface gravity was not found by studies carried out during the 1970s.)

Reduction in "G"

G here is the universal gravitation constant G - which affects the force of gravity experienced between two bodies. This is also something of an "exotic" explanation" - which, according to the theory would "cause expansion through the release of elastic compressional energy throughout the Earth." Maxlow reports that Carey rejected this proposal as the change in surface gravity would have, again, been excessive. Also, considering the limits of a change in G (which is, after all, meant to be a constant!) the magnitude of expansion would likely be too small to fit the available evidence. Carey also noted that the arguments for a reduction in G could not account for an exponential rate of Earth expansion. (We will mention this theory again later, however.)

Mass Increase by Plasma Transfer / LENR

In BPT section/chapter 12.1 James Maxlow shows the diagram below:

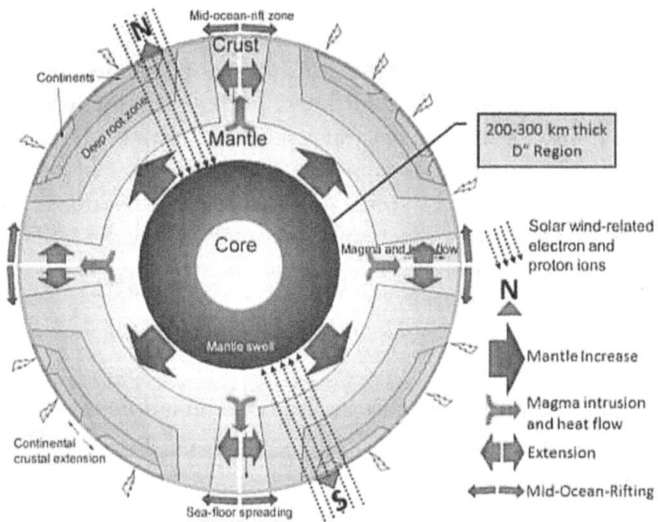

Maxlow's Caption: A schematic cross-section of the present-day Earth highlighting the influence of charged electrons and protons entering the Earth resulting in increase in mass and radius over time.

The basic idea here is that matter is "absorbed" from the solar wind and, somehow, transmuted to solid material in the core.

Maxlow quotes John B Eichler (a retired Physicist and Mathematician who studied and worked at the Illinois Institute of Technology)[165]. Maxlow writes:

The near-Earth observations presented here are based on a suggestion put forward by Eichler in 2011. Eichler posed the question, "Does plasma from the Sun cause the Earth to increase in size?" and by presenting a new argument based on known physical phenomena, he suggested that this might indeed be the case. Eichler elaborated with his statement that:

*"To assume that the Earth is gaining matter and that this may be due to nucleosynthesis within the Earth seems to fly in the face of conventional wisdom - and it does. Based on empirical geologic evidence which strongly indicates that… [an increase in Earth size] …is indeed valid, the task confronted is to formulate a viable mechanism whereby this occurs. In a plasma universe, the Earth is under constant bombardment from space, with all the necessary components to reconstitute matter from its component parts deep within the Earth not requiring theoretical constructs which have never been experimentally observed. The Earth, having a magnetic field strong enough to interact with impinging particles, gathers more than sufficient fundamental particles, namely electrons and protons, to account for a slow increase in matter internally over hundreds of millions of years. There is therefore no lack of component particles to create new matter deep within the body of the Earth. **The exact process by which this occurs is complex in nature and, like the interior of the Earth itself, involves speculation as to its dynamics.**

In BPT Section 12.2.1, Maxlow includes further details of Eichler's proposals:

It is envisaged that magnetically charged electrons and protons enter the Earth's magnetosphere and lower terrestrial layers primarily at the polar auroral zones and as random lightning strikes during electrical storms. These magnetically charged particles are further attracted by conduction to the strongly magnetic core-mantle region of the Earth. The elevated core-mantle temperatures and pressures present enable the particles to dissipate and recombine via nucleosynthesis as new matter within the upper core or lower mantle regions, in particular the 200 to 300 kilometres thick D" region located at the base of the mantle directly above the core-mantle boundary.

Maxlow suggests the D" region in the interior is of special interest because of other research by a German Geologist Professor Gerhard Kremp[166]:

This proposal also incorporates the observations of Kremp who, in 1992[167], suggested that new geophysical evidence indicates that the Earth has been growing rapidly in the past 200 million years. Kremp indicated that seismologists have located the existence of a zone, about 200 to 300 kilometres thick, located at the base of the mantle directly above the core-mantle boundary, designated the D" region.

That is, Maxlow identifies a region within the core where new mass might be being created by the process that Eichler describes. Maxlow then quotes other research related to heat flow within the outer core and suggests that Kremp concluded this heat flow from the inner to the outer core could be a recent development - related to the Earth's expansion.

Maxlow also proposes that the strength of the Earth's magnetic field may have changed over time and it is for this reason that the rate of expansion that he has established has also changed. That is to say that the strength of the Earth's magnetic field affects how much plasma or other charged particles would be trapped. However, he does not appear to discuss if this change in strength is recorded in the paleomagnetic data which he discusses at length elsewhere in his work.

Whilst the regular auroral displays at the poles prove that solar wind particles are indeed attracted to the Earth because of complex interactions between the Earth's and the solar magnetic field - and their electric fields, there is no clear description of exactly how these particles would then travel down into the depths of the mantle - or lower subterranean regions and combine to form solid matter. It does not really seem realistic to suggest that auroral or lighting events could play any significant role in transporting any significant mass deep into the Earth. Again, we have to explain the *exponential* rate of expansion - and Maxlow has not shown evidence of an exponential increase in lightning or auroral activity of the relevant time periods.

One might consider that some additional hydrogen and helium atoms might form somewhere in the Earth's atmosphere during lighting or auroral events, but how these atoms would then fuse to create heavier elements - such as carbon, silicon, iron and other metals - is not addressed.

Maxlow, to his credit, mentions Low Energy Nuclear Reactions (LENR) research - also known (incorrectly) as Cold Fusion, but does not go into any detail about transmutation of elements. This area has been researched and discussed in other papers[168], however, such as one published in 2002 by V.A. Kirkinskii, Yu. A. Novikov[169].

Neal Adams and Pair Production

When I interviewed Neal Adams in 2015[129], he spoke at some length about his theory that the Earth has "grown" (not "expanded")[170] and this is primarily the result of "pair production[171]." As I alluded to earlier, Neal has "his own spin" on aspects of quantum mechanics (such as those proposed to occur at the edge of a black hole, where Hawking Radiation[172] could be produced). Adams suggested that extra mass in the Earth's core has been created by this "pair production" process. This sounds quite similar to (though less detailed than) the "Plasma Transfer" process described by Maxlow. However, the main difference is that Adams suggests material is created in the core spontaneously, rather than being transported from the exterior, as Maxlow suggests.

Whilst Adams' idea is interesting, it is difficult to give support to it - because the pair production process that has been observed experimentally[173] involves penetrating radiation - like gamma rays - and such radiation is not likely to penetrate deeply enough below into the Earth's mantle to have the required effect. Also, what material and elements would be created by such a process? As with the proposed "Plasma Transfer" process, we would need a related process, operating on a large scale, to transmute the created matter into an appropriate range of elements, in the appropriate quantities.

Whole Earth Decompression Dynamics

An interesting website by J Marvin Herndon, PhD describes another theory which attempts to explain Earth expansion. This website does not seem to be that well known - and I came across it quite late in my research into the Earth expansion topic. Independent geophysicist researcher J. Marvin Herndon, president of the Transdyne Corp. (a scientific research and management company in San Diego, California), explains his theory in two short[174] YouTube videos[175]. Similarly, on his website "Nuclear Planet"[176], he provides the following useful summary and some illustrations

Early Earth Formation as a Jupiter-like Gas Giant

He writes:

*J. Marvin Herndon's concept of Earth originally having formed as a Jupiter-like gas giant leads to a new vision of Earth's internal composition, new geodynamics that correct and extend plate tectonics, powerful new energy sources, and georeactor magnetic field generation. In short, a whole new indivisible geoscience paradigm, securely anchored to the properties of matter. Figure at right, from left to right: 1) Earth condensing inside a giant gaseous protoplanet; 2) Fully formed gas-giant Earth; 3) Gases striped away by **T-Tauri solar wind;** 4) Ancient Compressed Earth (64% present diameter); 5) Present Earth; 6) Jupiter for size comparison.*

Herndon bases his idea on the "T-Tauri solar wind" phenomenon - a kind of giant solar flare - first observed by the Hubble Space Telescope in the late 1990s.

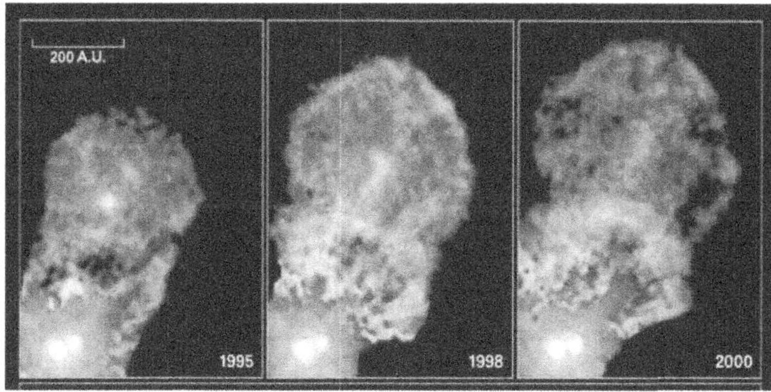

Hubble Space Telescope image of T-Tauri outburst from the binary XZ-Tauri - indicating movement of about 130 AU.[177]

We can see this giant flare has extended 13 billion miles in 5 years. Relating to this observation, then, in an article titled "J. Marvin Herndon's Whole-Earth Decompression Dynamics" he writes[178]:

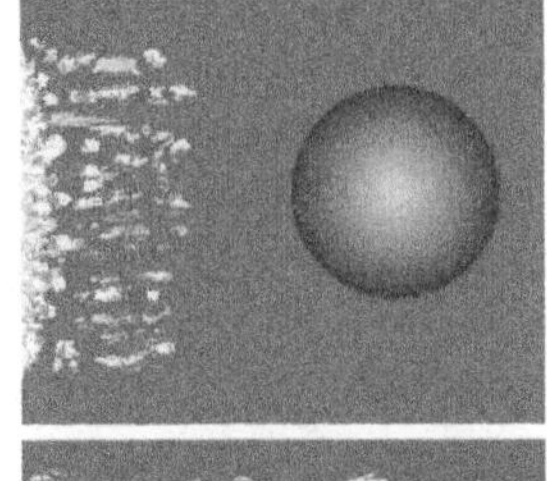

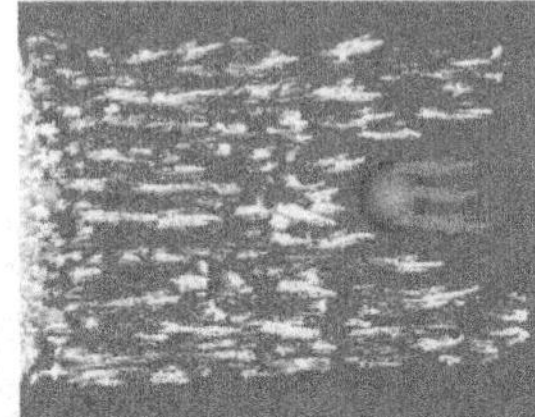

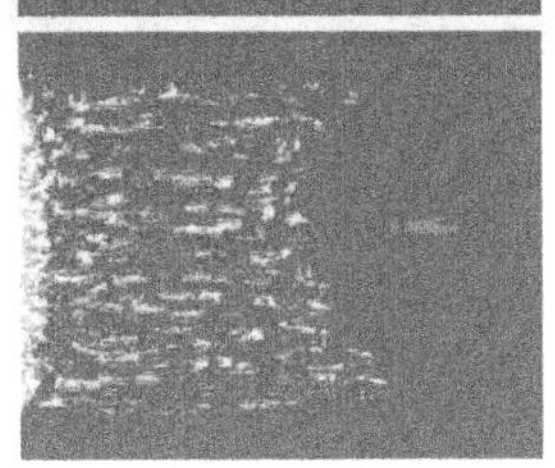

Decompression of the Earth may be seen as a direct consequence of the subsequent removal of hydrogen and other volatile constituents from the compressed kernel, presumably during the thermonuclear ignition of the Sun, as illustrated at left. After being stripped of such a great overburden, the Earth would rebound, tending toward a new hydrostatic equilibrium. Gravitational energy of compression, stored during the Jupiter-like proto-planetary stage, may be seen as the primary energy source for driving geotectonic activity, augmented to a much lesser extent by nuclear fission and radioactive decay energy.

Herndon has also posted an expanded version of this article in a paper called "A New Basis of Geoscience: Whole-Earth Decompression Dynamics"[179] (this does not seem to have been published in a recognised journal however).

It is presumed that Herndon assumes a constant mass Earth (once the much larger gaseous atmosphere had been stripped off). He never replied to the e-mail I sent asking him this question. As part of his discussion, Herndon, I think, makes a good case in refuting the notion of "mantle convection"[178] as being a major driving force of plate tectonics and continental motion:

When a fluid is heated from beneath, it expands becoming lighter, less dense, than the fluid above it. This top-heavy arrangement is unstable, so fluid motions result as the fluid attempts to restore stability. The top-heavy arrangement occurs because the temperature at the bottom is hotter than at the top. This is convection. Not only is the Earth's mantle not a fluid, but the weight of over-burden rock causes compression within the mantle, which increases with depth. Matter at the bottom of the mantle is about 62% more dense than at the top, as shown in the figure at right. Heating bottom-rock causes a miniscule increase in volume, hence miniscule decrease in density, much, much less than 1%. This is far, far too little to make the "parcel" of bottom-mantle light enough to float to the top, not enough to make the mantle top-heavy; the result is no mantle-convection at all. Moreover, the tacit assumption that the solid mantle behaves as an ideal gas with no viscous loss, i.e., adiabatic, is incorrect as evidenced by earthquakes at depths as great as 660 km.

In essence, Herndon cogently argues that you can't really have mantle convection *and* the occurrence of earthquakes at a depth where convection is supposedly also occurring.

Though Herndon's theory does not rely on added mass, it does not readily account for the evidence compiled by Maxlow regarding the exponential nature of the increase in radius. Herndon's suggestion that a single event stripped out the outer gaseous layers would lead to a conclusion that the expansion (or rather "decompression") started rapidly and he does not explain why the rate of expansion would increase exponentially.

Herndon seems to have made assumptions about the way the Earth has formed and has not explained why the general nature of the Earth now is different to the other inner planets. For example, Mercury and Venus are different from each other - and the Earth. Mars is also different to the Earth and the other inner planets - and though Neal Adams has argued Mars is also expanding (which might tie in with Herndon's argument), there is less available evidence for the expansion of Mars than is available from the study of the Earth.

Added Mass Theories Examined

In this section, and the following chapters, we will be including a number of calculations involving basic addition, subtraction, division, multiplication and squaring of numbers (multiplying them by themselves) and the use of the square root. We will also use what is called "scientific notation" or "standard form" to express some quantities. These calculations are, of course, used to support the assertions made, so that interested readers can check them. I know from experience that some readers simply balk when they see a "squared" or "square root" notation and think that they won't understand what is being shown. However, this cannot easily be avoided, other than the reader skipping over these calculations! Though this book is written for a lay reader and audience, I also wanted to ensure that those with the knowledge of how to do calculations will also be able to follow the reasoning used - and know that it has a sound basis.

In section (appendix) 27.3 of BPT, Maxlow includes some calculations relating to how the mass, density and surface gravity of the Earth have increased over time and includes the following figures for rates of increase:

Radius	22 mm/year
Circumference	140 mm/year
Surface Area	3.50 km²/year
Volume	11,000 km³/year
Mass	60×10^{12} tonnes/year
Surface Gravity	3.4×10^{-8} msec^{-2}/year

A key formula used here is the one for calculating surface gravity:

$$g = \frac{GM}{r^2}$$

G is the gravitational constant, a figure of 6.67×10^{-11}.

M is the mass, in kilogrammes, of the Earth - 5.972×10^{24} kg.

r is the (present) radius in kilometres - 6371km.

If you do this calculation, you will come out with a figure for "g" of $9.81 m/s^2$.

To explain the next step as simply as I can, we will look at a simpler calculation. Let us say the product of "GM" above (i.e. G x M) is 200. We say that "r" is 10 (for the sake of illustration). The value of r multiplied by itself (squared) is therefore 100.

Hence, the calculation becomes 200 divided by 100 - which of course is **2**. Now let us change the value of "r" to 5 instead of 10 - so r squared (r^2) would then be 5 multiplied by 5 to give us 25. This means our calculation would become 200 divided by 25 - which gives us **8.**

Hence, with a radius *r* that has been **halved**, the result of the same calculation is **four times** bigger than the original one.

This is a long-winded way of explaining what is known as "the inverse square law." That is, that in the case of measuring the surface gravity based on the radius of the Earth, if we were to reduce the radius to half the present value (to "set" the size of the Earth to what it was when it formed), then the surface gravity would be **four times** greater than it is at the moment, if the **mass** (M) did not appreciably change.

This means that for added mass theories to explain even a *constant* surface gravity, the *mass* of the Earth must increase *exponentially*. This is why Stephen Hurrell proposed that the Earth's mass must have *increased by a factor of about 8 since the time of the dinosaurs*. The mass has to increase *a lot* - because the radius has increased - and this factor is *squared* in the calculation used to calculate for surface gravity.

Using the added mass theory of Earth expansion alone (without explaining the constantly increasing rate of added mass), there would be no net gain in surface gravity. It seems very difficult to accept that the mass could have increased so much to cause gravity to be much stronger now than it was, when it was possible for mega flora and fauna to exist - 200 mya.

The Answer?

In the next chapter, we will study the evidence to support a theory developed by Peter Woodhead, a retired property developer from Lancashire, UK. Woodhead became interested in Earth Expansion research following his consideration of evidence discussed by fellow UK researcher Keith Hunter[180], relating to the early solar system[181]. As Peter Woodhead suggests:

> *...as a theory, added mass has been shown to have no merit, holds no appeal to the general public or scientific community and should be discarded. We are left with the need for a mechanism for Earth expansion that does explain the increase in surface gravity.*

This can be found in the next part of this book. Woodhead's theory is notable because it does *not rely* on added mass and it probably explains the exponential rate of Earth expansion. It does not rely on "exotic physics" to explain how matter can be created inside the Earth. It relies, primarily, on a mechanical process. However, the theory encourages us to reconsider how the force of gravity experienced at the Earth's surface is manifested — but we will try to show how this consideration leads us into another significant area of "alternative knowledge" research called "The Electric Universe."

97

11. A Mechanism for Earth Expansion

In the preceding chapters of this part of the book, I have laid out the evidence which, as far as I am concerned, proves that our planet has expanded since its formation. Additionally, I have briefly covered some of the evidence that the force of gravity experienced at the surface has also changed.

The current and following chapters in this book have essentially come about as a result of an earlier posting I made of the presentation I compiled about the "Hollow Earth" and "Expanding Earth" topics. Sometime after posting this "earthy" presentation online, I was contacted by a man called Peter Woodhead, who is based in the North West of England - he is a RichPlanet TV viewer[182]! He said to me he had an explanation as to why the Earth was expanding!

I went to see Peter in April[183] and October 2014[184] and we recorded two separate interviews relating to the ideas that he (and I) developed. He gave me all his diagrams, notes and calculations and I wrote them up and edited them and added some references and further diagrams. The first article about Woodhead's research was posted in June 2014[185] and the second was posted in April 2015[186]. As neither Peter nor myself are practicing scientists, we have never attempted to get any of the research published anywhere except on my own website and I would be the first to accept that more work would need to be done to bring the articles up to a level where they could be put into a scientific journal of some kind (even one which was run by interested academics who don't have any formal connection to a funding body or University). So, if any suitably experienced or qualified readers wanted to "take up the mantle" (pun intended), please do contact me at ad.johnson@ntlworld.com!

Formation of the Earth (again)

Let us first revisit the accepted model of the formation of our Earth and solar system. Picture in your mind the origin of our planet and you will no doubt conjure up images of fiery collisions and the moon and Earth being battered or even torn apart! Perhaps the truth is somewhat different. Also, we can again ask, what do we really know about our Earth's interior? We have been told that our planet has an iron core which is surrounded by a molten outer core and mantle. As we briefly discussed in chapter 6, we are told that this has been deduced from a study of seismic readings taken when earthquakes happen at various places around the globe[187]. However, we can also consider that seismic waves are not effectively or efficiently transmitted through solid / liquid /solid boundaries. Hence, if we accept that we have found viscous, molten rock some way below the Earth's crust, we may need to consider other explanations for certain seismic observations!

Another reason for the assumption of an iron core is related to the existence of our planet's magnetic field. This can also be questioned, perhaps, because when a bar magnet is heated sufficiently, it loses its magnetism.[188] Of course, as we mentioned in chapter 1, the standard explanation for the presence of the field is the "dynamo effect."

Later, we will consider, in some detail, the "Electric Universe" model and what the electrical interaction between the Earth and the Sun might be[189]. For the moment, let's just consider the Earth's rotation in the presence of an electric field (from the Sun) and consider the ferrous metals in the mantle and the Earth's crust. Perhaps this is all that is needed to induce the Earth's magnetic field - without recourse to an iron core.

So if we remove our preconceptions about an Iron Core, what should we replace them with? We can review some more recent observations of star systems where stars and planets are still forming and consider if these observations might apply to our own solar system.

Published on 18th July 2013, an article entitled "Snow in an infant solar system: A frosty landmark for planet and comet formation," discusses the work of astronomers at the European Southern Observatory (ESO)[190]:

> *A snow line has been imaged in a far-off infant solar system for the very first time. The snow line, located in the disc around the Sun-like star TW Hydrae, promises to tell us more about the formation of planets and comets, the factors that decide their composition, and the history of the Solar System.*

The article goes on to state:

> *Starting from the star and moving outwards, water (H_2O) is the first to freeze - forming the first snow line. Further out from the star, as the temperature drops, more exotic molecules can freeze and turn to "snow", such as carbon dioxide (CO_2), methane (CH_4), and carbon monoxide (CO).*

Hence, due to the different freezing points of different chemical compounds, different "snow lines" can be found at various distances from the star. It is understood that heavier elements in our system have been formed in supernovae (exploding star) events and these are present in the cloud of material that our solar system formed out of. In the case of TW Hydrae, we can assume a similar "cocktail" of elements are present. In a spinning accretion zone, the majority of the lightest elements (as in a centrifuge) accrete to the inner zone and the heavier elements to the outer reaches.

An artist's concept of the snow line in TW Hydrae showing water ice covered dust grains in the inner disc (4.5–30 astronomical units, blue)[191] and carbon monoxide ice covered grains in the outer disc (>30 astronomical units, green). The transition from blue to green marks the carbon monoxide snow line. The snow helps grains of dust to adhere to each other by providing a sticky coating, which is essential to the formation of planets and comets. Due to the different freezing points of different chemical compounds, different snow lines can be found at various distances from the star.

The Sun is principally comprised of hydrogen, which, according to the standard model of solar physics, fuses to form helium[192] - these 2 elements have the lowest atomic weights. So let us assume that Earth formed in our solar system's "water-snow zone". Let us further assume that in the sticky "snow" (as described by the ESO astronomers) were the rest of the elements that became our planet.

Heating an Icy Core

So what do we have? A lump of predominantly ice, at a few degrees above absolute zero, together with an accumulation of other elements created by the left-over stardust from a supernova.

What next? Well, having mopped up the majority of the available matter, the sphere of spinning matter which would become the Earth (pre-expansion) would be baked by the Sun. At this stage, with little atmosphere, Earth would be inhospitable to life. Below the newly forming crust, the interior would still be perhaps just a few degrees above absolute zero. We therefore propose that the Earth is then heated by a number of methods:

1) In the early stages, some heat from the Sun on the crust would travel by conduction and possibly convection through the crust and heat the interior.

2) Friction between the various materials which comprise the proto-Earth.

3) Subterranean heating from nuclear reactions - mainly radioactive decay (see below).[193]

4) Electric currents, induced by the Earth's movement in the Sun's electric field.[189]

Together, these heating processes would begin to slowly thaw the Earth's icy interior. Additionally, as an atmosphere formed around the Earth, it is possible that this also trapped additional heat from the Sun.

As the interior thaws, over millions of years, the developing Earth's magnetic field, might have a stirring effect on the "slush" which it turns into. That might cause further friction, resulting in additional heating of the core and the mantle.

An article called "Fixed-Earth and Expanding-Earth Theories - Time for a Paradigm Shift?" by David Noel (Revised 2005)[194] suggests how further heating could occur as a result of mechanical stresses or forces that appear during the Earth expansion itself (which, of course, we have been studying in the second part of this book):

Clearly the amount of frictional heat produced from such cubic-kilometre-sized crumpling and thrusting is immense in everyday terms, and fully sufficient to produce all the above geothermal effects, including melting to produce igneous rocks and heating to convert metamorphic ones. The pressures involved are also immense, sufficient to account for metamorphic processes usually attributed to deep burial in the Earth - as, for example, diamond formation.

Heating by Radioactive Decay

Over the last few years, several sources have suggested that radioactive elements may play a role in heating the Earth's interior. For example an article on UC Berkley News titled "Radioactive potassium may be major heat source in Earth's core"[195] states,

Radioactive potassium, uranium and thorium are thought to be the three main sources of heat in the Earth's interior, aside from that generated by the formation of the planet. Together, the heat keeps the mantle actively churning and the core generating a protective magnetic field.

Similarly, a July 2011 article in "Physics World" entitled "Radioactive decay accounts for half of Earth's heat"[193] states:

> *About 50% of the heat given off by the Earth is generated by the radioactive decay of elements such as uranium and thorium, and their decay products. That is the conclusion of an international team of physicists that has used the KamLAND detector in Japan to measure the flux of antineutrinos emanating from deep within the Earth.*

For a smaller sized Earth, the heating effect from radioactive decay would be more significant. Also, arguably, there would be more radioactive material - as it would not have decayed as much.

Liquid and then Gases Migrate to the Core

Over the next four billion years, the crustal heating would gradually heat the icy core. First any frozen gasses, hydrogen, nitrogen, oxygen and other less abundant gaseous elements would be created. Some of these lighter gases/elements would migrate to the very *centre* of our planet - and some would come to the surface of the crust. I say this following the revelation of a simple experiment on the International Space Station - where air bubbles are injected into a floating drop/small sphere of water[196] - the bubbles migrate to the centre[197].

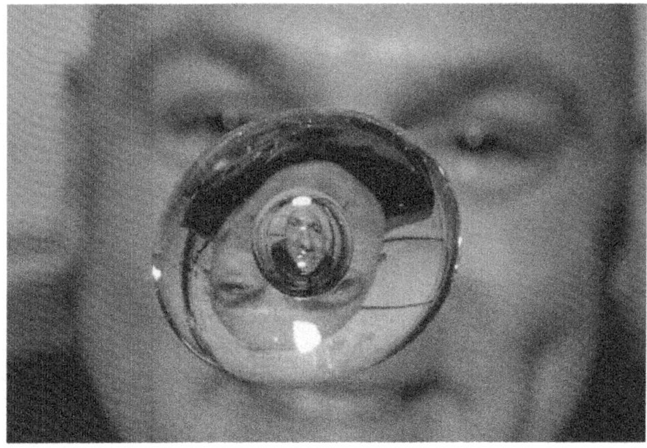

An air bubble, trapped inside a water droplet, on the International Space Station

In the developing Earth, we can posit that the gases and lighter materials will then travel to the core, where the area of *least pressure* and zero gravity exist! Yes, this is *the opposite* of what conventional thinking suggests, but is based on the observations in the water droplet experiment mentioned above.

So, to summarise - we are suggesting that we have a *core of ice* in the Earth, with a crust of heavier elements around it. The ice would begin to melt and we would end up with a core filled with *water*.

"Steam-Powered Earth Expansion!"

Anyone familiar with steam power and pressure vessels will tell you that without regulation of steam pressure, even the thickest steel vessel will explode. It is worth noting that, at current atmospheric pressure, water when heated to steam expands by a factor of about 1600 times.[198]

Having considered the formation and development of the early Earth, to the point where the liquid core starts to boil, we can now consider our planet to be a "slowly exploding pressure vessel." In this scenario, the crust is going to crack along the lines of least resistance i.e. "the continental margins". The water (predominantly) was forced through these cracks and manifested as high-pressure steam - along with other heated materials. We propose that over the following period of tens of millions of years, this venting created our ever-deepening oceans. We have called this "Steam Powered Expansion" or "Gas Powered Expansion." Hence, we will use the abbreviation "GPEE" later in this book.

Despite this venting of pressure, it was not enough to restrict the expanding core, the result being the stretching of the newly forming ocean bed at the later to be known as "spreading ridges". The expansion would then proceed as discussed earlier in this book.

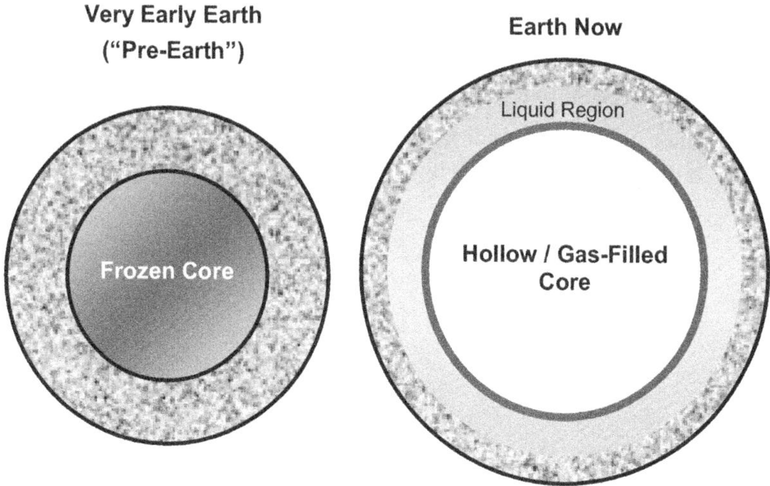

Proposed Ancient and Present Structure of the Earth

One might suggest that if the core was originally ice, which then melted, then perhaps the Earth shrank slightly when all the ice had melted - because, remember, ice is less dense than water and floats! However, this situation would depend on exactly how the heating took place - because it would certainly have been possible that some of the water reached boiling point before all the ice had melted. (For example, consider placing or resting a red-hot iron rod or poker on top of a slab of ice and see if any steam is ever created...)

Of course, many physicists would immediately reject the idea that water could boil down at the core of the Earth - they would say that the pressure down there would raise the boiling point so much that the water probably would not boil… However, we can answer this point, at least in part, by noting that a lot of water comes out of volcanos - and volcanic vents in the ocean.

Calculating the Amount of Water

In the following pages, we attempt to calculate the volume of ice/water in the core and we will then compare this to the estimated volume of the Earth's oceans - just to see "where that takes us." This is all to help us quantify, in some way, *how* the expansion has proceeded and whether the idea of a water/gas powered expansion is at all feasible.

We will now attempt to quantify likely volumes and sizes of the core. Let us first note our current Earth's radius, which is approximately 6370km. We also note the following facts and figures:

- The combined land area of all continents[199] is about 149 million km².
- The area of the Earth's continental shelf[200] is about 29 million km².
- The combined total area is about 178 million km².

Since the rest of our planet's area consists of basalt formed at the mid ocean ridges, we know this part did not exist in "pre-expansion" times[201].

Pre-Expansion Radius

Given a surface area of 178 million km², we can calculate Earth's pre-expansion radius to be about 3,764km.

$$a = 4\pi r^2$$

$$r = \sqrt{\frac{a}{4\pi}}$$

$$r = \sqrt{\frac{1.78 \times 10^8}{4\pi}}$$

$$r = \underline{3763.6km}$$

As noted above, the average radius of the Earth today is about 6,370 km. Earth's pre-expansion radius was, therefore, roughly 59% of today's radius. This figure agrees quite well with the value of 3500km, calculated by Klaus Vogel[202]:

Mr. Klaus Vogel (EEE, March 1980) - It was by some reference to the possibility of Earth's expansion that I became aware of the theory of Earth Expansion. Instantly, it occurred to me that A. Wegener's "Pangaea" could have been a completely closed surface of a much smaller Earth. Attempting to join the pieces led to astounding results. It led to a "Continental Crust Sphere" of a __diameter__ of approximately 7000 kilometres.

However, Maxlow suggests an even smaller "primordial" Earth radius of 1700km.[203] Here, we will use the 3763km figure, for the purposes of argument.

Pre-Expansion Volume and "Gas Volume"

Using the formula

$$v = \frac{4}{3}\pi r^3$$

We can calculate the Earth's pre-expansion volume of

$$v = \frac{4}{3}\pi \times 3763^3$$

This gives a figure of 223,307,760,801 km³ (approx. 2.23 x 10¹¹ km³). We now calculate our Earth's current volume:

$$v = \frac{4}{3}\pi \times 6370^3$$

This gives a figure of 1,082,696,932,430 km³ (approx. 1.08 x 10¹² km³). We now subtract the pre-Earth volume from the current Earth volume:

$$1{,}082{,}696{,}932{,}430 - 223{,}307{,}760{,}801 = 859{,}389{,}171{,}629 \text{ km}^3$$

$$= 8.6 \times 10^{11} \text{ km}^3$$

Rounding down, this gives a "gas volume" (v_g) of approx. 860 billion km³. Using this figure, we can now calculate the approximate gas radius, r_g:

$$v_g = \frac{4}{3}\pi r^3$$

$$r_g = \sqrt[3]{\frac{3v}{4}\pi}$$

$$r_g = \sqrt[3]{\frac{3 \times 8.6 \times 10^{11}}{4}\pi}$$

$$r_g = 5900\text{km}$$

The gas radius of approximately <u>5,900 km</u> is the radius of the Earth's "hollow" (gas-filled) core, at this point in time. From this, we determine that the thickness of the crust, mantle and water reservoir is

$$6370 - 5900 = \underline{470km}$$

Please note the **470km** figure - we will return to this later!

Solid and Liquid Inside the Earth

Billions of years of warming of the "slushy" (i.e. ice and water mix) interior eventually turn it into a liquid, then a super-heated liquid. Heat will also have accumulated in the now superheated upper mantle, which has, in turn, melted the ice core. We are now getting to the point where the liquid core will start to come to the boil - and something has to give…!

We have suggested that originally, the Earth was made up of a combination of ice and other solid materials. We don't know the original proportion of ice (water) to solid materials that was present after the Earth formed. The ice has now melted and the resulting water has started to boil, but not all of the water has boiled yet. At present therefore, we have a mixture of solid crust, liquid - and gas.

This liquid region will be made up of a proportion of water and a proportion of other molten materials. The ratio of water to other liquids and gasses is open to speculation.

We have already calculated the gas volume inside the Earth, so we will now attempt to calculate the volume of liquid in the following way:

1) We will calculate the volume of solid material from the assumed thickness of the Earth's continental crust and oceanic crust.
2) We will use the estimated figure for **the total volume of the oceans**. We are therefore suggesting that most of the water in the current oceans has come from *inside* the planet (more on this later).
3) Knowing the total volume of the Earth, we will then subtract the figures in (1) and (2) and also subtract the gas volume. This should allow us to calculate the volume of liquid below the mantle.

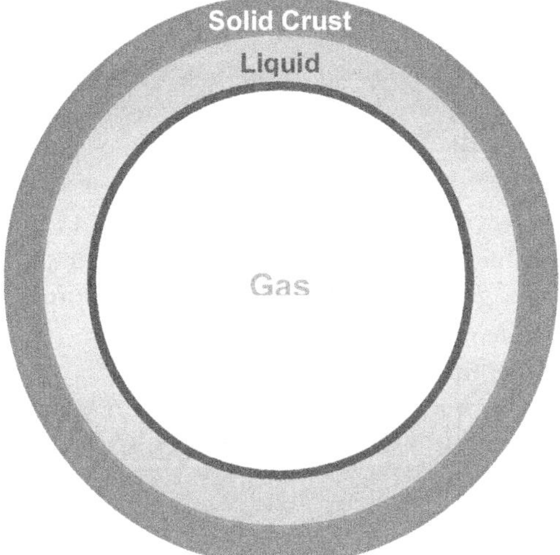

Current state of the inside of the Earth - proportion of the Solid Crust/Liquid/Gas amounts is a calculated guess.

We will now attempt to calculate the average thickness of the crust over the whole surface of the Earth. Our Earth's crust is estimated to be 40km thick[204]. The oceanic crust is thought to be about 6km thick[205] and average ocean depth is 4km[206].

The surface area of the Earth is given by $a = 4\pi r^2$:
$$a = 4\pi \times 6370^2$$
$$a = \underline{5.1 \times 10^8 \text{ km}^2}$$

This figure can be written as 510 million km². As shown earlier, the total surface area of the continental shelf and all continents is 178 million km². Hence, the area of the oceanic crust is

$$510 - 178 = \underline{332 \text{ million km}^2}$$

We will now calculate an overall average thickness using the proportion of oceanic crust area to continental crust area.

$$av.\ crust\ thickness = 6 \times \frac{332}{510} + 40 \times \frac{178}{510} = \underline{17.9 km}$$

If we round the average to 18km, the radius of the "gas + liquid" sphere (the liquid includes the mantle) is then given by

$$6370 - 18 = \underline{6352 \text{ km}}$$

We can also calculate the depth/thickness of the liquid layer by subtracting the gas radius (5900km) from the "gas + liquid" radius:

$$6352 - 5900 = 452 \text{ km}$$

Deduced General Structure of Interior - An Inner Crust?

Although we have no easy way of determining the proportions of water/other liquid in the liquid portion, we can consider there will be a **second, solid layer between the mantle and the water**. The hottest part of the mantle will be somewhere in between the bottom of the outer crust and the water layer boundary. As we go down through the mantle, towards the water layer, it should get cooler. At the boundary between mantle and liquid zone, in exchange for the heating effect on the liquid (water), a corresponding cooling of the mantle will create a solidified layer - an Inner Crust. We can therefore envisage a structure like this:

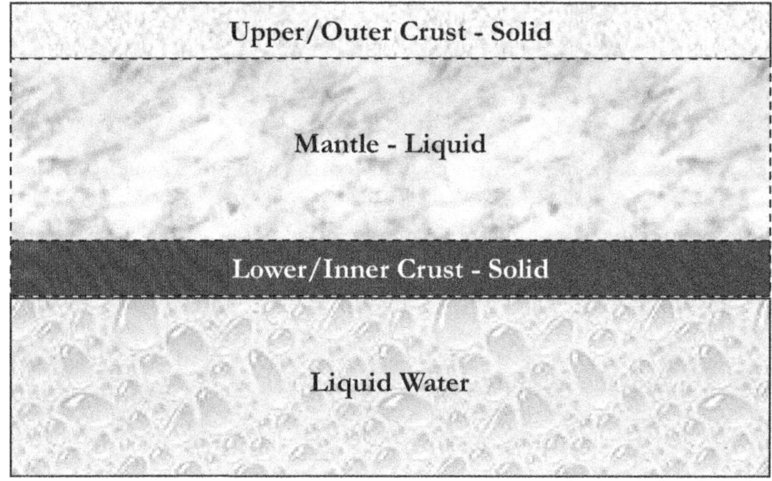

Here are 3 possible scenarios with different proportions of liquid and solid material between the crust and the gas core. We will then simply divide up the 452km liquid thickness by these proportions. We will assume the inner/lower crust is of similar thickness to the upper/outer crust - say, 14km.

Combined Liquid Thickness (Mantle + Liquid Water) = 452 - 14 = <u>438km</u> (we will round this to 440km)

Layer ↓	Solid/Water Ratio →	30% / 70%	50% / 50%	70% / 30%
Mantle (km thickness)		132	220	308
Water (km thickness)		308	220	132

Suggested Cross section of Earth 70% Liquid Non-Water and 30% water

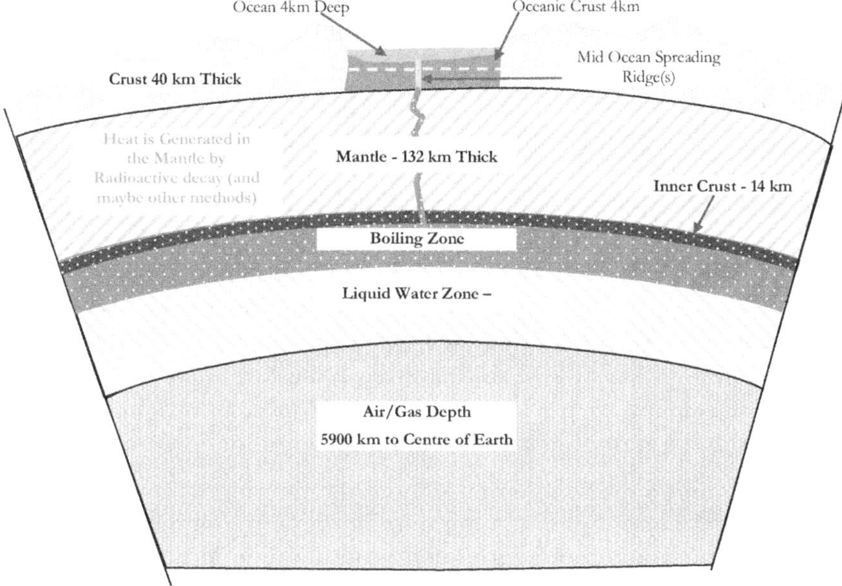

Notes

1. At the mid ocean spreading ridges, hydrothermal vents have recorded temperatures of up to about 400 degrees Celsius and at 3000m depth pressures of 300 atmospheres![207] It is reasonable to assume slightly greater pressure on the mantle side.
2. Lava reaching the surface during eruptions has a temperature of 700-1250 Celsius[208] therefore we must assume a higher value for the Mantle itself.
3. Beneath the inner crust is the "boiling zone". High pressure steam is created, some of which migrates to the inner hollow core, adding to the "Steam Powered Earth Expansion".

Volume of Water in the Inner Earth - and the Oceans

If we take the 70% water value, the thickness of the water layer would be 308km. The volume of water underneath the inner crust is then given by subtracting the volume of the gas sphere from the "gas + water" sphere:

The radius to the outer edge of the water sphere is given by:

$$5900 + 308 = \underline{6208km}$$

$$v_{water} = \frac{4}{3}\pi(6208^3 - 5900^3)$$

$$v_{water} = 141,781,565,722$$

$$v_{water} = \underline{1.4 \times 10^{11} km^3}$$

It has been estimated that our oceans currently hold a volume of 1,335,000,000 km³[209] . This represents, based on the calculations above, just <u>less than 1%</u> of what may still be beneath the mantle.

$$proportion\ of\ ocean = \frac{1.3 \times 10^9}{1.4 \times 10^{11}} = 0.009$$

This means that there is more than enough water down there!

Water/Solid Ratio

With this calculation done, we could also make a less extreme argument in relation to the "mechanical part" of the Earth expansion process. Even if there is actually less than 10% of the water/ice in the core in the model we proposed earlier in this chapter, there would still be more than enough water to enable the steam-powered Earth expansion.

Any Old Iron?

Whilst in later sections/chapters, we will present further evidence in support of the GPEE model, there are some obvious problems. For example, we have "discarded" the idea of a molten iron core and "replaced" it with ice/water/gas. This means we now have a "missing mass" problem because the specific gravity (density) of iron is about 7.9 (it is 7.9 times as dense as water). Perhaps this will cause many to reject our idea as "ridiculous." However, in a later chapter, we will make further arguments about how the force of gravity might operate on a planetary scale. This may then affect how mass is measured - as, for planetary bodies, the mass is inferred from the measurement of velocities and accelerations of those bodies. The amount of inferred mass is therefore related to the assumed force of gravity at any given location - which is what we will be re-considering later.

12. GPEE, Oceans, Oil and Moons

In this chapter we will explore evidence that the majority of the water in our oceans came from inside the Earth.

As a starting point, we can state that most geologists/Earth scientists agree that more than about 200 mya, there were no deep oceans - only shallow seas.[210] We can therefore immediately suggest that there is a possibility that the deep oceans formed round about the same time that the exponential Earth expansion, as documented by Maxlow, commenced.

Maxlow on Ocean Origins

In his BPT book, in section/chapter 16.3, Dr James Maxlow goes into some detail about the rise and fall of sea levels and writes:

> *The variation in coastal outlines on small Earth models shows that the total volume of ocean water in the past was very much less than what it is now. This contrasts with conventional studies where the total volume of ocean waters is considered constant, or near constant, over time. Small Earth modelling also shows that the volume of ocean water has been increasing steadily since Archaean times and most prominently since the post- Permian crustal breakup and opening of the modern oceans.*

But where did all this extra water come from? Maxlow continues thus:

> *This post-Permian increase in volume of new water has occurred in conjunction with intrusion of new volcanic seafloor crust along the global network of mid-ocean-ridge spreading zones. The new water, plus accompanying atmospheric gases, represents escaped volatile elements which occur naturally within the crystal lattices of all molten volcanic rocks. Petrological studies show that up to fifteen to twenty percent of the weight of a new volcanic rock may comprise entrapped fluid and gaseous elements. Once the volcanic rocks are intruded as lava near the Earth's surface, these volatile elements are then expelled, or **boiled off**, during formation of new surface lava or seafloor volcanic crust.*

Maxlow then later writes, in chapter/section 23.1:

> *The primitive atmosphere and hydrosphere - the combined mass of all water found on Earth - was considered by Lambert in 1982 to have been formed largely from elements and molecules degassed from the Earth's interior and subsequently modified by physical, chemical, and biological processes. Rubey proposed as early as 1975 that degassing - the removal of dissolved gases from liquids [inclusive of molten magma] - has been a continuous or recurrent process, which is still occurring today.*

He then quotes William W Rubey[211] thus:

> *"the whole of the waters of the oceans have been exhaled from the interior of the Earth, not as a primordial process, but slowly, progressively and continuously throughout geological time."*

Maxlow then notes:

Studies of melted igneous rocks carried out since the 1970s and 1980s have shown that the solubility of water in melted rocks increases with increasing pressure and temperature until a maximum value is reached in the mantle. Quoted examples range from 14 to 21 percent by weight of water dissolved in volcanic rocks at temperatures varying between 1,000 to 1,200 degrees Celsius accompanied by high pressures.

Underground "Reservoirs of Water" Discovered

Maxlow's referenced statements and comments are in agreement with postings/articles I found around the same time I originally posted the "GPEE" articles. For example, an article in "Nature", dated 12 March 2014 entitled "Tiny diamond impurity reveals water riches of deep Earth."[212] The article states:

A microscopic crystal of a mineral never before seen in a terrestrial rock holds clues to the presence of vast quantities of water deep in Earth's mantle.

Serendipitously, on the same day that I posted a video about our GPEE research, [183] an article was published in the New Scientist 12th June 2014 edition, in an article entitled "Massive 'ocean' discovered towards Earth's core"[213]. From this article, we can note:

The huge size of the reservoir throws new light on the origin of Earth's water. Some geologists think water arrived in comets as they struck the planet, but the new discovery supports an alternative idea that the oceans gradually oozed out of the interior of the early Earth.

"It's good evidence the Earth's water came from within," says Steven Jacobsen of North-western University in Evanston, Illinois. The hidden water could also act as a buffer for the oceans on the surface, explaining why they have stayed the same size for millions of years.

This research received considerable exposure[214] in other publications[215], although research from approximately 7 years earlier[216] was not mentioned.

An article entitled "3-D seismic model of vast water reservoir revealed" from February 2007[217] reports:

*A seismologist at Washington University in St. Louis has made the first 3-D model of seismic wave damping - diminishing - deep in the Earth's mantle and has revealed the existence of an underground water reservoir at least the volume of the Arctic Ocean. **It is the first evidence for water existing in the Earth's deep mantle.***

Michael E. Wysession, Ph.D., Washington University professor of Earth and planetary sciences in Arts & Sciences, working with former graduate student Jesse Lawrence (now at the University of California, San Diego), analysed 80,000 shear waves from more than 600,000 seismograms and found a large area in Earth's lower mantle beneath eastern Asia where water is damping out, or attenuating, seismic waves from earthquakes.

Previous predictions calculated that a cold ocean slab sinking into the Earth at 1,200 to 1,400 kilometres beneath the surface would release water in the rock that would escape the rock and rise up to a region above it, but this was never previously observed.

An Ocean Towards the Earth's Core?

The image shown below appeared[218] on a blog called "Renaissance Universal" following the 2014 postings about a "subterranean ocean."

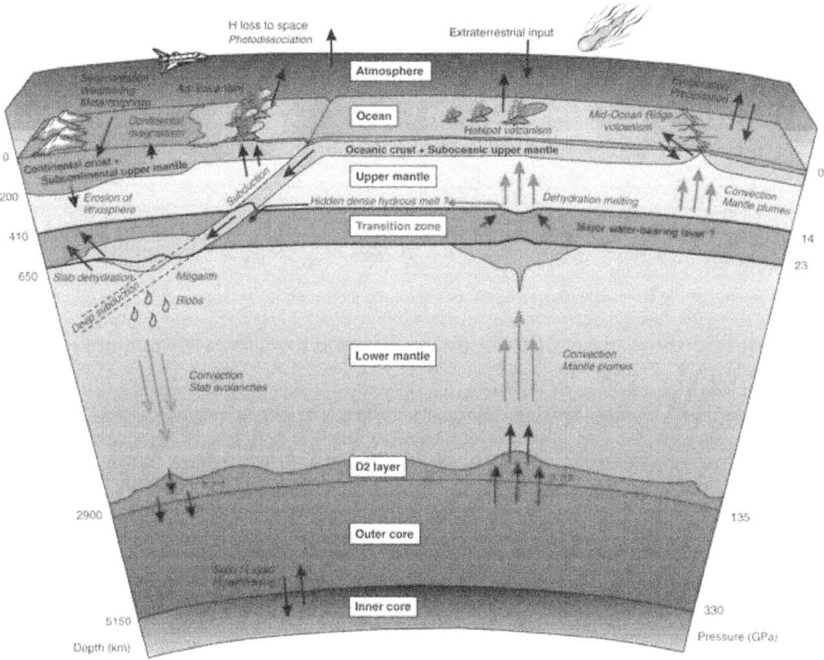

This model suggests much less water would be present than in the Gas-Powered Earth Expansion (GPEE) model. However, a small amount of water does not seem to negate the GPEE model altogether - as has been previously stated, steam expands to many times the volume of the water it is formed from[198], so it can be calculated that enough water will still be available based on the suggested depth and thickness of the transition zone shown above.

We note on the diagram above that the "water bearing layer" – the "Transition Zone" is at **410km** which seems to correspond quite well with the figure of **470km** that we calculated in the "Pre-Expansion Volume and Gas Volume" section of the previous chapter.

Ringwoodite and "Slush"

Earlier, it was proposed that there was a layer of what we called "slush" - some kind of material which was not quite water, but neither was it ice. We also calculated the volume of water in the inner Earth, based on the figures we had suggested then. In the 2014 "New Scientist" article referenced above, we find the following picture and caption:

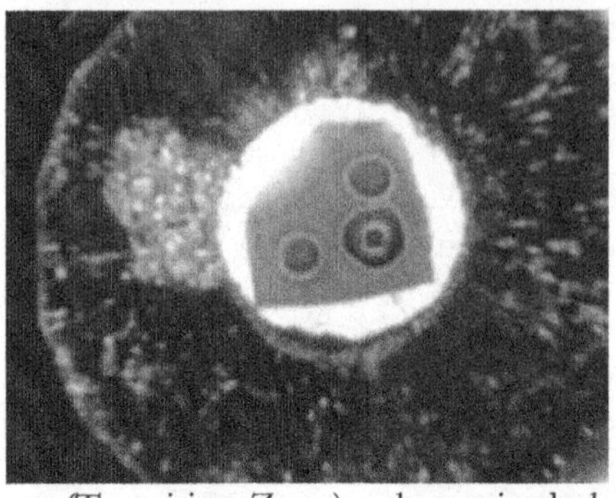

Blue lagoon: this crystal of blue ringwoodite is being crushed in a lab experiment. The orange circles are regions that have had their water squeezed out of them

Now, with the new/additional information about the proposed Ringwoodite layer, which is (presumably) contained somewhere in the "water bearing" layer (Transition Zone) - shown in dark blue - in the diagram above, we can improve some of our earlier calculations.

We can suggest from our GPEE model that the Ringwoodite layer might be undergoing an expansion of its own, as the 5% or so of water is heated and released into the surrounding layers.

The depth and thickness of the Transition Zone, as shown above, are 410km and 240km respectively. The radius of the "bounding spheres" would then be 6370 - 410 = 5,960 and 6370-650 = 5,720km. Assuming a uniform thickness of this zone around the Earth, the volume of this zone would be given by:

$$= \frac{4}{3}\pi(5960^3 - 5720^3)$$

$$= \frac{4}{3}\pi(211{,}708{,}736{,}000 - 187{,}149{,}248{,}000)$$

$$= \frac{4}{3}\pi \times 24{,}559{,}488{,}000 \text{ km}^3$$

$$= 1.03 \times 10^{11} \text{km}^3$$

It is stated that Ringwoodite's water content is between 1.5% and 2% by weight.[219] Since ringwoodite has a specific gravity of 3.9g cm^3 then the water contained by volume will be somewhere between 3.9×1.5 = 5.85 and 3.9×2 = 7.8. These 2 figures are then averaged to 6.845% by volume. Hence the volume of water in this layer alone would be:

0.068 x 1.03 x 10^{11} = 7.02 x 10^9 km^3

It has been estimated that our oceans currently hold a volume of 1,335,000,000 km^2[203] - the volume calculated above is approximately **5 times** this figure. This therefore lends credence to the idea that the oceans could have come from underneath the present mantle.

One thought to consider regarding the origin and formation of the oceans is the age of the bed of the Mediterranean Sea (shown on the "rainbow" ocean floor map[221]). Could the first out-welling of basalt and water have occurred with "venting" from below the mantle in this area? Could that venting have been initiated by a meteoric or cometary impact? Perhaps further study of geological features may reveal more about this possibility.

Some additional questions… why does the ringwoodite contain the water in the first place? How did it get "sandwiched" between the hot mantle material and the even hotter molten core? It should long since have vented and evaporated into the atmosphere, should it not?

Origin of the Deep Oceans

The previous discussions of large volumes of water being present under the mantle led us, logically, to look at two related questions in relation to the current deep oceans that we have. Where did the *volume* of water come from and where did all the salt come from?

If you try to research these questions online, there seems to be a dearth of relevant information. One posting I came across, from 2005 (now archived away!)[222] was made by Dr Bill White, Professor of Earth and Atmospheric Sciences at Cornell University[223].

> *Q: Where does the salt in the sea come from?*
>
> *A: …scientists today know that this is only part of the answer - for example there is hardly any chlorine or sulphur in most rocks so those elements couldn't come from weathering.* **They come from volcanoes - which spew out gases containing sulphur and chlorine as well as lava."**

The only other information I could easily find was in an unpublished book called "In the Beginning by Immanuel Velikovsky[224]." Before I quote that, it's worth noting that between the 1950s and 1970s, Velikovsky - a true polymath - was something of a "thorn in the side" of the Academic Establishment[225], as he published several books which challenged many mainstream conclusions about planetary sciences and cosmology. The most well-known of these is probably "Worlds in Collision."[226] Whilst I am not sure what to make of Velikovsky's claims about Venus once being a part of Jupiter, and other claims Velikovsky made about Saturn, we have to accept that he made certain predictions about Venus and Jupiter, which turned out to be correct. For example, he predicted Jupiter would emit radio waves.

Carl Sagan (left) seemed to have been tasked with "debunking" Velikovsky's (right) claims[227]

Velikovsky's "In the Beginning" book is available in an online archive, he has a chapter about the origin of the oceans[228]. Here we can read the following.

> *A part of the salts could be traced to the washing of lands and the floor of the seas; chlorine is known also to be discharged by volcanoes, but to account for the chlorine locked in the seas, volcanic eruptions, whether on land or under the surface of the seas, needed to have taken place on an unimaginable scale - actually, it was figured out, on an impossible scale. Thus it was acknowledged that the provenance of chlorine in the salts of the seas is a problem unsolved.*
>
> *Paleontological research makes it rather apparent that marine animals in some early age were more closely related to fresh-water fauna; in other words, the salinity of the oceans increased markedly at some age in the past.*
>
> *The proportion of salts in the rivers is very different from their proportion in the seas. River water has many carbonates (80 percent of the salts), fewer sulphates (13 percent) and still fewer chlorides (7 percent). Sea water has many chlorides (89 percent), fewer sulphates (10 percent) and only a few carbonates (0.2 percent). The comparison of these figures makes it clear that rivers cannot be made responsible for most of the salts of the seas. Therefore it is also obvious that there is no proper way of calculating the age of the Earth by comparing the amount of salts in the seas with the annual discharge by the rivers; the most that can be done in this respect is to calculate the rich amount of carbonates in the rivers in their relation to the relatively poor concentration of these salts in the seas; but then there will be no explanation for the rich concentration of chlorides in the seas in comparison with their poor concentration in the rivers.*

Velikovsky suggests the salt water came from Saturn, although I would find this hard to accept myself (considering other areas covered in this book).

Though the source of the information about rivers is unreferenced, we know from experience that river water around the world is not salty! Salinity is only really significant where river flow is affected by the tides (i.e. in estuaries, or where water flows into rivers from them). It is hardly worth saying that we only have freshwater species of plants and animals in a river until all the way down its course, until we find tidal waters. Indeed, while we are on this subject, we can note the amazing metamorphosis of Salmon from a salt water fish to a fresh water fish…[229]

Another point we can observe is that how these enormous rifts in the Earth's crust have been filled so perfectly with saltwater.

The main alternative theory about the origin of the abundance of water is that it comes from comets. However, it has been known for some time that comet tails have water rich in deuterium (heavy water), while the oceans have much less deuterium in their water[230].

Let us then consider where most of our salt has come from - salt domes which are found all around the world and they can be *miles* in diameter.[231] Salt domes can be up to 6000 feet below the surface of the Earth.

Also, on a NOAA page asking "Why is the ocean salty?"[232] we can read:

> *By some estimates, if the salt in the ocean could be removed and spread evenly over the Earth's land surface it would form a layer more than 500 feet thick, about the height of a 40-story office building.*

That, to me, sounds like an enormous amount of material to be washed down off the land and into the ocean - especially when we consider the lack of salinity in most (all?) rivers….

Coal and Abiotic Oil

As well as Earth expansion, we can thank our "gas/steam" powered core for the creation of oil, in a process known about for well over a hundred years "abiosis." Methane, forced through Earth's crust by the immense pressure from Earth's core, in combination with steam is transformed to the "black gold".

Contrary to popular belief, no organic life has ever existed at the depth that most oil reserves[233] are said to become "fossilised."[234] It has also been discovered that some oil wells previously considered "dry" have subsequently been replenished (though it is argued this is through seepage due to a release of pressure when reserves have been extracted[235]).

Again one might argue that the conditions for formation of coal could be created by the steam and pressure derived from a gas-powered Earth expansion process.

Gases Venting from the Surface of Other Bodies

Having shown a likely source of gases (particularly water) venting from the interior of planets, we can now look elsewhere in the solar system to see if there are any other places where this is happening.

Images of Jupiter's moon Europa taken by the Hubble Telescope and published on 12/12/2013 have been found to show water being vented into space through fissures in its frozen crust.[236]

Further images and data have been published - since as early as 2001. Published in Oct 2011, images sent back by the Cassini probe show venting on the moon Enceladus. (It seems similar jets were also photographed in 2010.)[237]

Enceladus is "a mirror-like image of Earth" - with seas of liquid methane and mountains, dunes and shorelines made up not of rock but frozen water ice! Titan and Enceladus could have formed in a methane accretion zone[238] - further out from our Sun, just as in the case of the disk around TW Hydrae (as we discussed in chapter 11).

Neptune's moon Triton is also suspected to have water.[239]

There are additional articles about subterranean water on other moons in our solar system. For example an article on Reuters from Oct 16, 2014 entitled "Saturn moon may have 'life-friendly' underground ocean: scientists"[240] reports:

> *CAPE CANAVERAL Fla. (Reuters) - Saturn's battered moon Mimas may have a thin global ocean buried miles beneath its icy surface, raising the prospect of another "life-friendly" habitat in the solar system,*

In relation to the Jovian moon Ganymede[241], an article dated 12 Mar, 2015 on NASA.GOV titled "NASA's Hubble Observations Suggest Underground Ocean on Jupiter's Largest Moon"[242] states:

> *NASA's Hubble Space Telescope has the best evidence yet for an underground saltwater ocean on Ganymede, Jupiter's largest moon. The subterranean ocean is thought to have more water than all the water on Earth's surface.*

In a "Nature.com" article dated 26 July 2001 - entitled "Callisto's watery secret"[243] the moon is described thus:

> *One of Jupiter's largest moons, Callisto, may hold watery secrets beneath its surface, suggests a new analysis. The satellite's icy crust may be the planetary equivalent of a blanket, insulating an underground ocean.*

Do these instances lend support to our theory of an ice core in planetary bodies/moons which liquifies/melts over thousands or millions of years?

Celestial Bodies "Ringing Like Bells"?

Earthquakes

It has been determined that most earthquakes occur at depths up to 40 miles - all within the mantle. However, earthquakes have been detected to a depth of 430 miles – approximately where the bottom of the ringwoodite layer has been said to be. About 3% are "deep" earthquakes, 12% are "intermediate" and 85% are "shallow."[244]

The GPEE theory also therefore could explain this situation - since the crust is brittle and the transition zone, ringwoodite with liquid water within will also be brittle, the zone between (mostly magma) is viscous.

It has also sometimes been said that the Earth can "ring like a bell" when an earthquake happens[245]. Does this imply, then, that the Earth really does have a shell-like structure - with a large void at the centre? One quote reads[246] "The Earth rings like a bell after a large earthquake - the lowest ring tone is E flat in the 20th octave below middle C (Source: F Press and R Siever, Earth, New York: Freeman, 1986, p 467)."

Another quote is[247]

> *Seismic waves from the biggest earthquakes (over magnitude 8.3) can bounce around inside the Earth for up to a month. This makes the Earth "ring". However, you need special instruments to hear the ring because the tone is very low - about 1 cycle per hour. Compare this with the 256 cycles per second of middle C on the piano.*

Though seismologists would argue this effect happens because of the difference in nature between the inner and outer core of the Earth etc, the comparison of "shell structure" to a bell seems to hold at least some merit. This might also add weight to the suggestion that the Earth does indeed have a physical "resonant frequency" - as is perhaps implied by the quotes above.

Moonquakes?

It has been alleged by NASA that seismometers were left on the moon by the Apollo Astronauts and that these have been used to detect moonquakes[248]. Though there is plenty of evidence to show NASA's claims of landing men on the moon are not truthful, it is possible that instrumentation has been remotely landed there instead. That said, I can include the following text from a NASA article about the alleged Apollo 15 landing[249]:

> *Knowledge of Lunar Interior Structure. Like Earth, the Moon has a crust, mantle, and core. The lunar crust is rich in the mineral plagioclase and has an average crustal thickness of 60-70 kilometres, which is about 3 times the average crustal thickness on Earth. The lunar mantle lies between the crust and the core and consists mostly of the minerals olivine and pyroxene. The core is probably composed mostly of iron and sulphur and extends from the centre of the Moon out to a radius of no more than 450 kilometres; i.e., the core radius is less than 25% of the Moon's radius, which is quite small. In*

Though planetary scientists may object, we can suggest, that like other bodies orbiting the sun, the moon could also have a gas-filled centre.

Formation of the Earth and the Moon

Though this remains a mystery, most agree the moon and the Earth formed at the same time (however, we have to note that a smaller number argue that the moon arrived much later, with some people going so far as to say the moon is manufactured - a giant space station even[250]).

For the moment, we will suggest the Earth and moon were originally two spinning globes, moving in a figure of eight formation, with opposite directions of rotation. Whilst remaining cold, their fast rotation would continue - for perhaps 4 billion years or more - all the while their cores being heated by induced currents from the Sun's electric field and radioactive heating in the crust (shell). During expansion, conservation of angular momentum would mean the rotation of each body would slow down. The viscosity of the moon's interior would be greatly influenced by the Earth's gravity and would be ever more drawn to the Earth, like an inner tide, slowing the moon's rotation - eventually to the "tidal locked" situation we have now.

Could the tidal locking lead to an inner anomaly in the thickness of the moon's shell? This may then lead to a different force of gravity on the Earth-facing side of the moon compared to the far side.

13. GPEE and Gravity

Conservation of Angular Momentum

Going back in time, Earth's diameter would be progressively smaller. As a consequence of this, the Earth would have rotated more rapidly (consider an ice skater drawing in their arms while spinning). This would mean the Earth's day was shorter and consequently there would be many more days in a prehistoric Earth's year.

Some calculations were done to determine the Earth's angular momentum, based on the current 24-hour day. From this, the approximate length of the Earth's day in the time of the dinosaurs can be calculated. Similarly, the equatorial centripetal force can be calculated, but this is essentially negligible. These figures are shown below.

mya	Day Length (Hours)	Equatorial Centripetal Force (N/kg)
300	8.38	0.16
250	9.09	0.14
200	9.86	0.13
150	10.95	0.11
100	13.07	0.08
50	17.03	0.06
10	21.29	0.04
0	24.00	0.03

These figures do not really help us to find an explanation for the gigantism of prehistoric flora and fauna.

Surface Gravity Considerations

As we noted in chapter 9, the dinosaurs inhabited our planet during early expansion and grew to an impossible size (based on the current surface gravity). On page 81 of Stephen Hurrell's "Dinosaurs and the Expanding Earth" book [251], we find this graph:

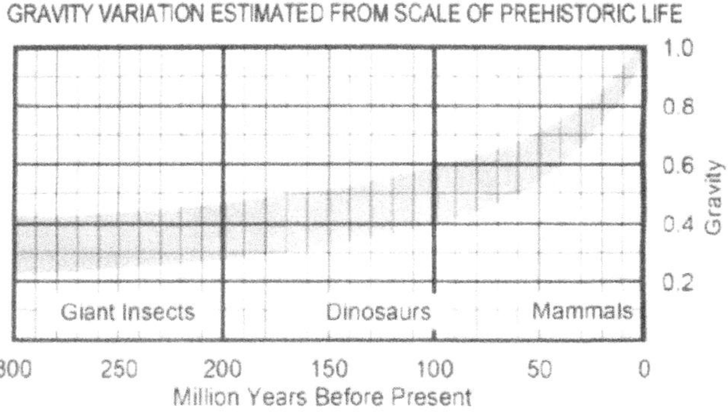

Please note the exponential curve of the ratio animal size/gravity. Hurrell proposes, therefore, that the surface force of gravity was between about 30% and 50% of its current value between 300 and 100 mya. What would cause this change?

Considering a new "Centre of Gravity"

Let's review our proposed structure for the Earth and then consider an idea that goes against most current thinking about how gravity is "meant to work."

We contend that Earth's mass now is not significantly more than that pre-expansion. If we consider that our pre-Earth's outer region was predominantly rock/heavy elements (pictured left), then its "Centre of Gravity" or "Centre of Mass-Force" in relation to a body on the surface, would *not* be at the centre of the Earth. Yes, we realise that **this is absolutely not how Newtonian Gravitation is supposed to work… but please read on!**

What we propose is that the reason for the increased force of surface gravity as the expansion proceeded is that **the mass distribution** beneath the surface of the Earth changed - which caused the centre of gravity (mass-force) *to move away from the centre of the Earth* and get closer to the surface - **as the non-gas layers get thinner.** Again, this is therefore contrary to the situation of a solid sphere where the centre of gravity is at the centre of the sphere!

As we did earlier, let us consider again the formula for the acceleration (attraction) due to gravity at the Earth's surface, g:

$$g = \frac{GM}{r_c^{\,2}}$$

M is the mass of the Earth and r_c is the radius at any given time. Simplifying the figures, for the sake of illustration, we suggest that if the centre of gravity is located at $r_c = 1000$km from the Earth's centre 200 mya. Hence, the denominator of the fraction above becomes, say, 1000 x 1000 = 1 million.

However, for the Earth as it is now, the centre of gravity is located, say 500km below the surface. The denominator in the equation then becomes 500×500 = 250,000. In simple terms, then, this would mean that force (acceleration actually) of a mass 200 mya at the surface would be ¼ of the force now.

Intuitively, when you look at a diagram showing an object in contact at the surface with the shell, it seems unlikely that the mass of the far hemisphere can exert the same force on the object as the near hemisphere. To establish where this CMF might be for the expanding Earth, a spreadsheet was developed. We realise we have simplified the scenario and the maths for the calculation, but again, please remain open to the idea being proposed.

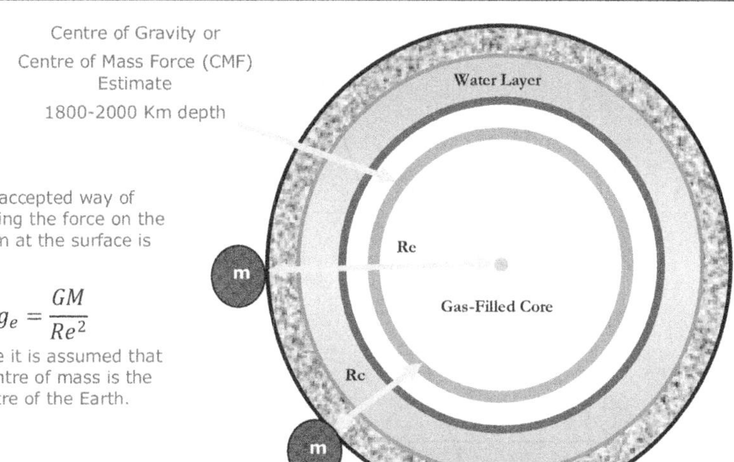

With the above equation, it is assumed that the centre of mass is at the centre of the Earth. R is the radius of the Earth - **which is the same as the distance to the assumed centre of mass**.

With a hollow or gas-filled shell, the centre of mass attracting a mass on the surface, would **not** be at the centre of the Earth - the mass of shell underneath "m" in the diagram above would be **closer** to the mass on the surface. Hence, the effective R would be smaller than the radius of the Earth (Rc < Re).

Surface Gravity and Centre of Mass Force

Developing these ideas further, we consider the gas-filled Earth being split into a number of segments, and consider the force of gravitational attraction on a mass at the surface. In the diagram below, each segment exerts a force on masses on the surface. The horizontal forces cancel each other out, but the vertical forces are additive. This is a kind of 2D-visualization. An enhanced 3D visualization is shown below.

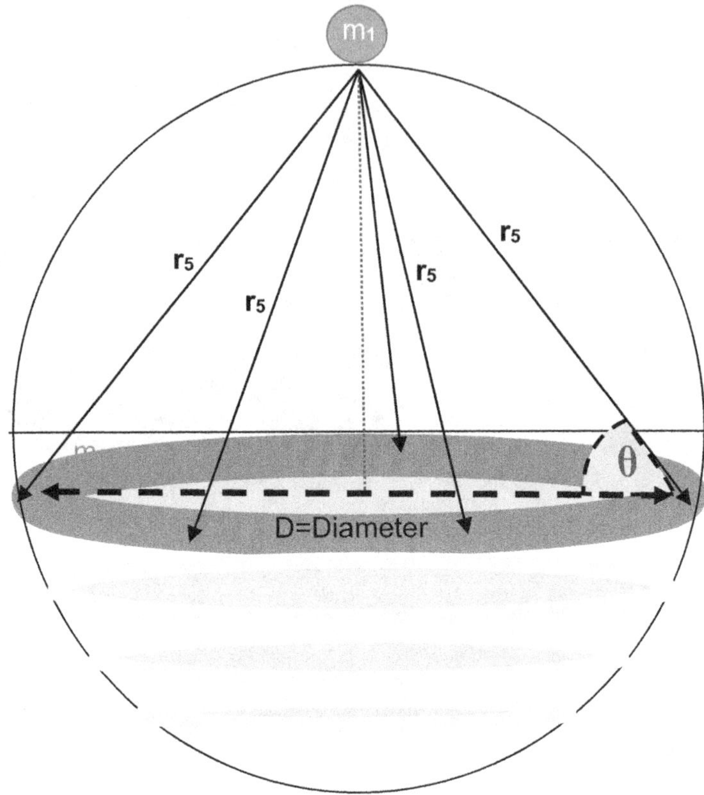

m₁ is a mass on the surface, which is attracted by the mass in all of the mass of the Earth

m₂ is a ring/shell. The volume of the shell is given by the area multiplied by the thickness of the ring and is therefore proportional to the mass. The vertical force is proportional to sin θ where θ is the angle of the line of force measured perpendicular to the equatorial diameter, D above. The forces exerted would need to be summed by integration (as in calculus).

In relation to this area of study and argument, a series of relevant questions and comments appeared in a Maths/Physics forum in May 2009, prompted by a poster named "geistkiesel"[252]

> *Where, and how, does the Newton Shell theorem (NST) place the CMF at the sphere COM? It doesn't. The question of locating the CMF is not discussed! The **NST model** begins with a ring on the sphere of differential mass dM oriented perpendicular to m and centred on r. By summing the force for each dM on each ring then integrating over the surface of the sphere, the total force, F = GmM/r^2 results, which says nothing, absolutely nothing, regarding the location of the CMF.*

Another poster (unsurprisingly) responded:

> *the result of the integration is:* $F_r = \dfrac{GMm}{r^2}$

Further discussion ensued, but posters did not seem to accept the validity of
the question. We now move on to the "NST model," mentioned above.

Newton's Shell Theorem (NST)

Some well-informed people will immediately point out that "the gravitational
force will still act as if it is at the centre of the Earth - even if the Earth is
hollow or gas-filled!" They will point to examples like this one[253]:

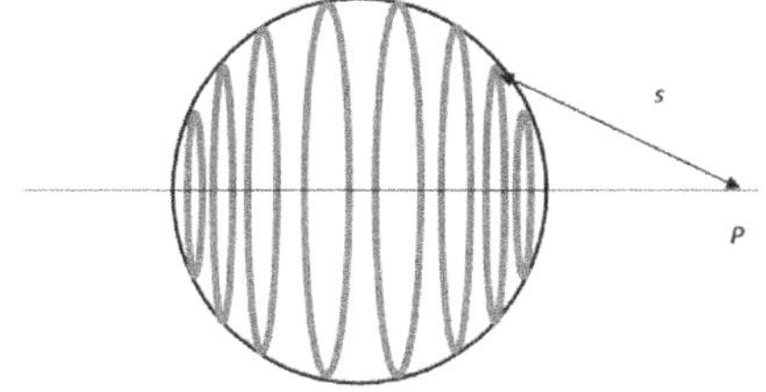

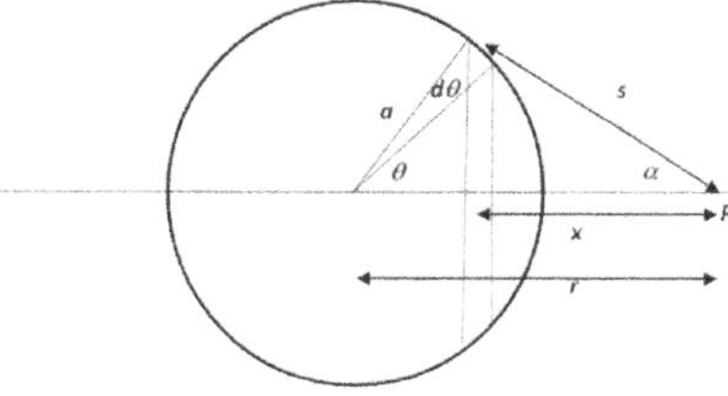

| Such a shell can be envisioned as a stack of rings. | Field Outside a Massive Spherical Shell |

It is hardly surprising then that, in researching a method for calculating the
CMF more accurately, we became aware of Newton's Shell Theorem, which is
stated thus[254]:

*Newton's Shell Theorem states essentially two things, and has a very
important consequence. First of all, it says that the gravitational field outside
a spherical shell having total mass M is the same as if the entire mass M is
concentrated at its centre (centre of mass). Secondly, it says that for the
same sphere the gravitational field inside the spherical shell is identically 0.*

This is explained quite well in a YouTube video entitled "Universal
Gravitation - Shell Theorem"[255]. Although the Shell Theorem is a
mathematical representation of how the forces involved *should* produce certain
predictable results, it is an "idealised" model - and if what we cover in later
sections is correct - at least partially - the precise results one should get from
Shell Theorem calculations may not match reality.

For example, if "the force always acts from the centre," then would it not be
the case that there should be no gravitational anomalies as measured from the
surface? We know that such anomalies are used to find oil wells or other areas
of geological interest, so we know that measurements of surface gravity are
neither uniform, nor straightforward.

Similarly, some people have compared Newton's Shell Theorem and Gauss's
law - and some physicists have even done work on "Gauss's Law for
Gravity,"[256] thus:

*In physics, Gauss's law for gravity, also known as Gauss's flux theorem for
gravity, is a law of physics that is essentially equivalent to Newton's law of*

CMF Proposals and the True Mass of the Earth?

Peter Woodhead has revised a number of the figures he calculated since we first published our web articles in 2014 and 2015 and, by considering the pre-expansion Earth radius, the current radius and the proposed current density of crustal rock, Woodhead argues we should agree that the Earth's mass can only be a fraction of the accepted figure. This is explained in more detail below, though it might take a while to "get your head around it!"

Woodhead developed two basic diagrams, showing some of the figures he worked with. These are shown below.

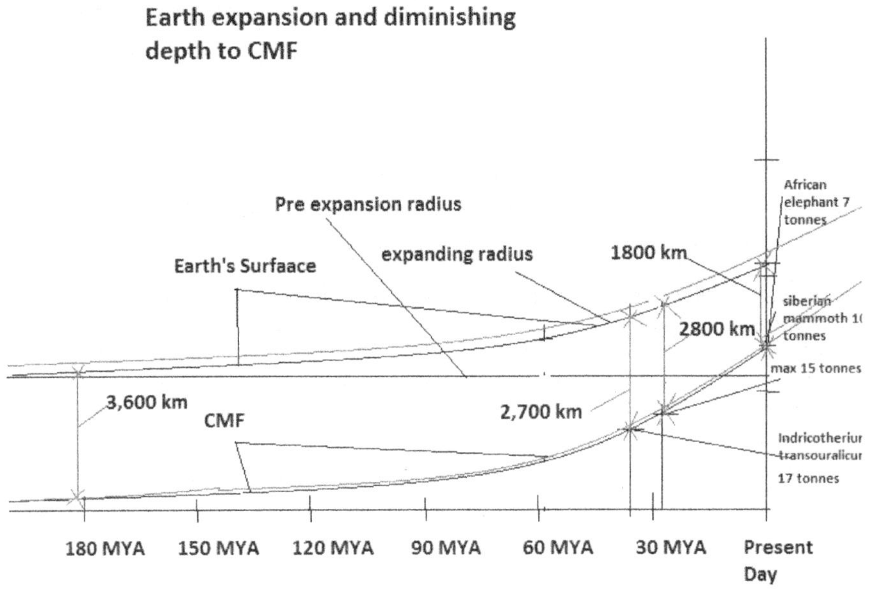

Curves showing Earth's expansion and change in the depth of the Centre of Mass Force

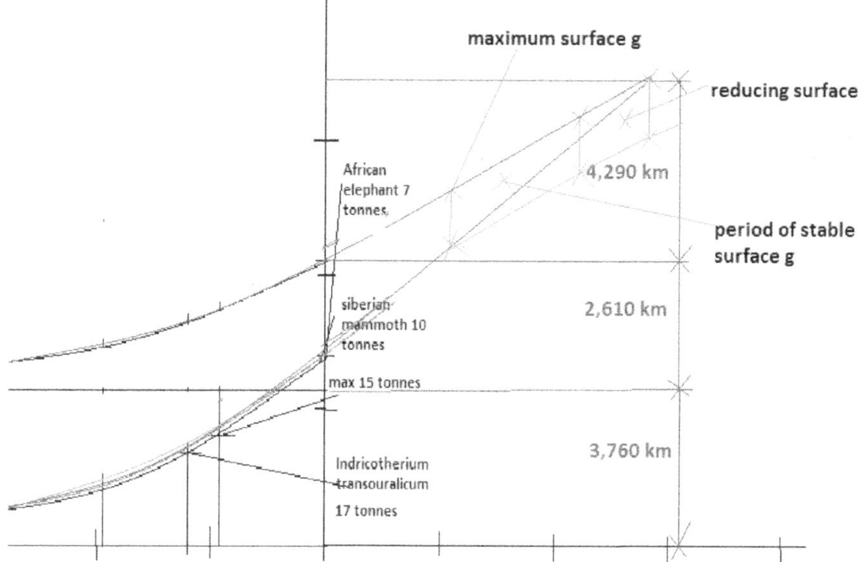

A more recent set of values from the last few million years.

Woodhead states that the current CMF is about **1800 km** below the surface. In order to reconsider what the Earth's mass might be, we first calculate the volume of the pre-expansion Earth thus:

$$v = \frac{4}{3}\pi r^3$$

We can calculate the Earth's pre-expansion volume of

$$v = \frac{4}{3}\pi \times 3763^3$$

This gives a figure of 223,307,760,801 km³ (approx. 2.23 x 10¹¹ km³). We now calculate the mass of the original Earth, based on the readily available information that the rocky shell of our planet is deemed to have a density between 2.6-2.7 kg/litre.

Woodhead then suggests we consider that we had a 30% volume of water (ice) in the original Earth and 70% was rock - of similar or the same density as the accepted crust density is now. Hence, the average density of the whole Earth originally would be:

$$d_{avg} = 0.7 \times 2.65 + 0.3 \times 1 = 2.155 \; kg/l$$

Using this density, we multiply by the original volume of the Earth, in litres (which is 10^{12} or 1000 billion times the volume in km³):

$$m_e = 2.155 \times 2.23 \times 10^{11} \times 10^{12} = 4.81 \times 10^{23} \, kg$$

As we mentioned in chapter 10, the *accepted* mass, in kilogrammes, of the Earth is 5.972×10^{24} kg.

The figure calculated by Woodhead, based on the current crustal density, the original Earth radius and the suggested ice/rock mix, is then compared to the accepted mass, thus:

$$\frac{4.81 \times 10^{23}}{5.972 \times 10^{24}} = 0.08$$

This then is just 8% of the assumed current mass! Woodhead argues that the acceleration due to gravity comes about not solely as a result of the mass, but *where that mass is*. Hence, as the CMF moves outwards as the Earth expands, the force of gravity changes.

Maintaining this line of reasoning, Woodhead further argues that if we use this re-calculated mass to calculate g for the pre-expansion Earth, we would then have much weaker gravity. In other words, he argues that to calculate the surface gravity now, we use the current accepted mass, but to calculate the surface gravity for pre-expansion, we must use the re-calculated (smaller) mass and the pre-expansion radius.

So, remembering that

$$g = \frac{GM}{r^2}$$

the divisor for calculating the *current* surface gravity is given by :

$$\frac{1}{r^2} = \frac{1}{6370^2} = 2.46 \times 10^{-8}$$

The divisor for calculating the *pre-expansion* surface gravity is given by :

$$\frac{1}{3673^2} = 7.08 \times 10^{-8}$$

Hence, just using the difference in radius in the force of gravity - g - calculation (and ignoring the recalculated mass), would mean that g was 7.08 / 2.46 or 2.88 times *stronger* on the pre-expansion Earth. However, including the recalculated mass we would have g then only 2.88 x 0.08 = 0.23 - or about ¼ of what it is today.

In summary, then Woodhead argues, gravity at the **pre-expansion** Earth's surface would be roughly 25% of that today. In other words, he is suggesting that the force of gravity at the surface is affected, to some degree, by the material the Earth is made of, and it is affected by the way the mass is distributed throughout the core and the mantle. Woodhead also argues that there was almost no expansion prior to the date given for the oldest Ocean Floor - about 180 Million Years.

With these considerations, then, we created some diagrams (<u>not</u> drawn to scale) to attempt to show a cross section of the Earth's structure as it expanded, to give some kind of illustration of the ideas and figures above.

Again, this goes very much against Newtonian physics, but in a later chapter, we will look at a way we might be able to explain the sorts of figures we are using. Maybe our explanation is some way off, and we would welcome sensible suggestions on how to improve it.

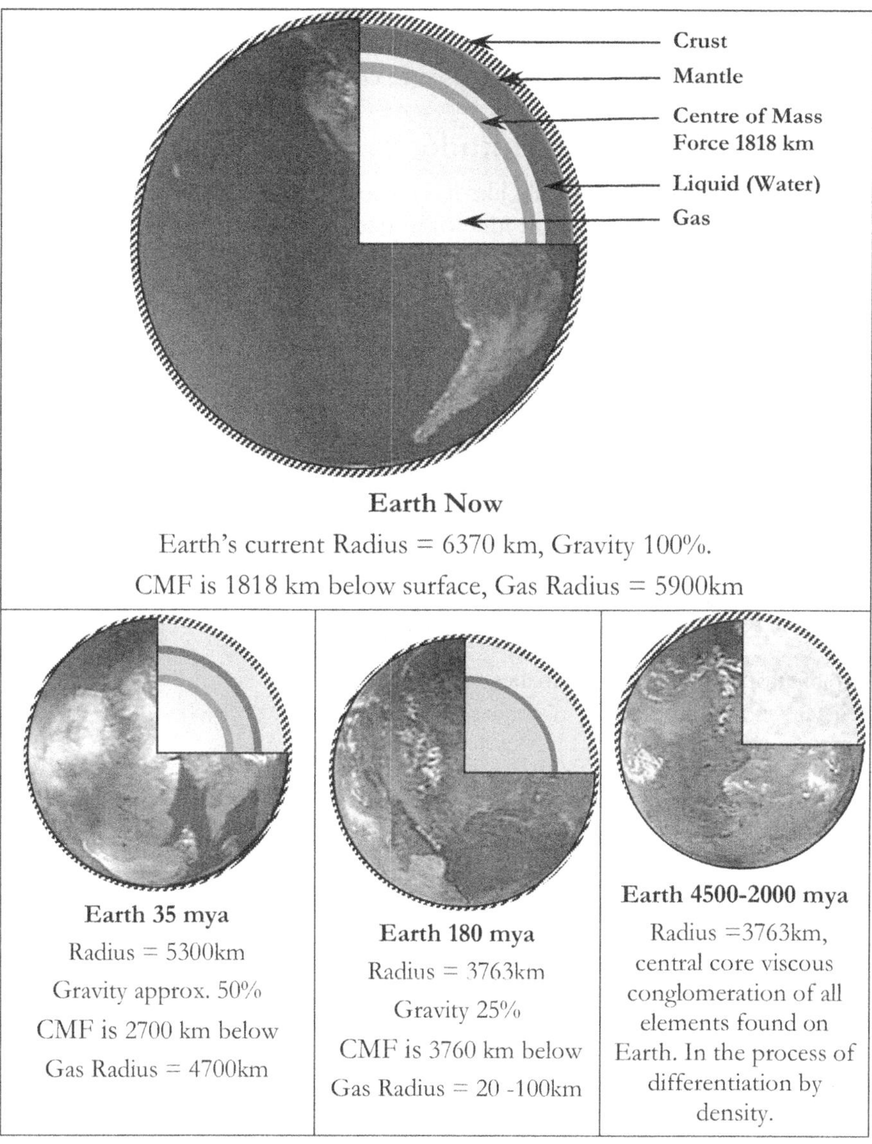

Earth Now

Earth's current Radius = 6370 km, Gravity 100%.

CMF is 1818 km below surface, Gas Radius = 5900km

Earth 35 mya

Radius = 5300km

Gravity approx. 50%

CMF is 2700 km below

Gas Radius = 4700km

Earth 180 mya

Radius = 3763km

Gravity 25%

CMF is 3760 km below

Gas Radius = 20 -100km

Earth 4500-2000 mya

Radius =3763km, central core viscous conglomeration of all elements found on Earth. In the process of differentiation by density.

Missing Mass and Dark Matter

Cosmologists have said that much of the Universe is composed of dark matter and dark energy[257] (and they can't be proved wrong, because we cannot easily see these "dark" components). Woodhead has proposed that large masses may only be about 10% of the mass we currently assume them to be. Woodhead's argument then brings us to consider that the so-called "cosmological missing mass" problem, which led to the need for the "invention" of dark matter, dark energy (and, perhaps, to a lesser extent, black holes) can now be "solved". We don't need to invent missing mass - because the force of gravity is affected, much more, by the way mass is distributed, not just by how much there is. In other words, there is already enough visible mass to explain the observed effects of gravity. This is a similar conclusion to the one argued by proponents of the Electric Universe model (discussed in chapter 14).

Gravity Anomalies at Altitude

In relation to our idea, we would like to consider the force of gravity when on a high mountain, one might consider it can be affected by 2 factors.

a) The force of gravity would be slightly **reduced** by being further away from the centre of the Earth

b) The force of gravity would be slightly **increased** by having "dense hard rock" beneath you.

Several web pages allow the force of gravity at altitude to be calculated[258] (all other factors being equal). Using the summit of Everest at 9,000 metres as an example, the page calculates the force of gravity will be reduced by 0.28%. This figure is essentially calculated as follows:

$$\frac{(earth's\ radius\ at\ sea\ level)^2}{(earth's\ radius\ at\ altitude)^2} = \frac{6370^2}{(6370+9)^2} = \frac{40576900}{40691641} = 0.9972$$
$$= 0.28\%\ reduced\ gravity$$

Explanations for gravity anomalies vary - for example, a study called "Geophysical anomalies along strike of the Southern Appalachian Piedmont"[259] suggests:

> *A regional trend in the Bouguer field is determined for an observed eastward decrease in crustal thickness based on seismic refraction measurements. A comparison of the calculated regional with surface boundaries shows that in places the crustal thinning occurs more than 50 kilometres west of the Inner Piedmont-Charlotte belt transition.*

Could this "crustal thinning" be a result of the gas-filled structure of the Earth? The differences in crust thickness would, arguably, have a much more significant effect on the local force of gravity, due to the reduced distance from the surface to the centre of mass force.

In August 2013, New Scientist published an article entitled "Gravity map reveals Earth's extremes"[260]. The article notes some extremes of gravitational acceleration. For example a reduction of about 0.5% (based on the accepted average acceleration due to gravity of 9.81 ms^{-2}) has been measured at the summit of mount Nevado Huascaran Peru in the Andes at 7,000 m - i.e.[261].

> *The Nevado Huascarán summit (Peru) with an estimated acceleration of 9.76392 ms^{-2}*

Using the afore-mentioned webpage[258], we would expect to find the acceleration due to gravity to be, 9.7814 ms^{-2} -a reduction of 0.32%

If we were to suggest this was the result of the CMF being at 2200km depth, we can calculate the expected change as follows.

$$\frac{(\text{CMF radius at sea level})^2}{(\text{CMF radius } + \text{ altitude})^2} = \frac{2020^2}{(2020 + 7)^2} = 0.9931$$
$$= 0.69\% \text{ reduced gravity}$$

This study identifies the places with the highest and lowest gravity acceleration:

A candidate location for Earth's maximum gravity acceleration was identified - outside the SRTM area, based on GGE-only - in the Arctic sea with an estimated 9.83366 m s^{-2}. This suggests a variation range (peak-to-peak variation) for gravity accelerations on Earth of about ~0.07 m s^{-2}, or 0.7 %, which is about 40 % larger than the variation range of 0.5 % implied by standard models based on a rotating mass-ellipsoid (gravity accelerations are 9.7803 m s^{-2} (equator) 9.8322 m s^{-2} (poles) on the mass-ellipsoid, cf. Moritz [2000]). So far such a simplified model is also used by the Committee on Data for Science and Technology (CODATA) to estimate the variation range in free-fall acceleration on Earth [Mohr and Taylor, 2005]. However, due to the inhomogeneous structure of Earth, presence of topographic masses, and decay of gravity with height the actual variations in free-fall accelerations are ~40% larger at the Earth's surface.

Could some of these anomalies be explicable by the CMF being in a position other than the core of the Earth? If we take the highest and lowest stated figures for g, we can explain the total variation measured fits well with the calculation made above i.e.:

$$9.83 \text{ (Arctic Ocean)} - 9.76 \text{ (Nevado)} = 0.07 \text{ ms}^{-2} \text{ variation}$$
$$0.07/9.81 = 0.0071 = 0.7\%$$

In a paper entitled "Specific Gravity Field and Deep Crustal Structure of the 'Himalayas East Structural Knot'"[262], a number of Gravity readings are shown in Figs 5, 6 and 7 (Fig 7 is shown below).

These and similar studies tend to explain these anomalies by inferring that beneath the mountain areas were areas of "less dense substrate", also that under the low altitude areas lay "much denser substrate". We would argue that the reductions are more significant because the CMF is closer to the surface than is accepted/used in the calculations. Fellow researcher Fredrik Nygaard (whose research is discussed in the next chapter) suggests:

If gravity changes with altitude in a different manner than what the shell theorem predicts, then Newton was definitely wrong. Mono-pole gravity should behave exactly as predicted by the shell theorem. Gravity with a directional/radial/dipole element will not conform to the shell theorem.

CHINESE JOURNAL OF GEOPHYSICS   Vol.49, No.4, 2006, pp: 932~940

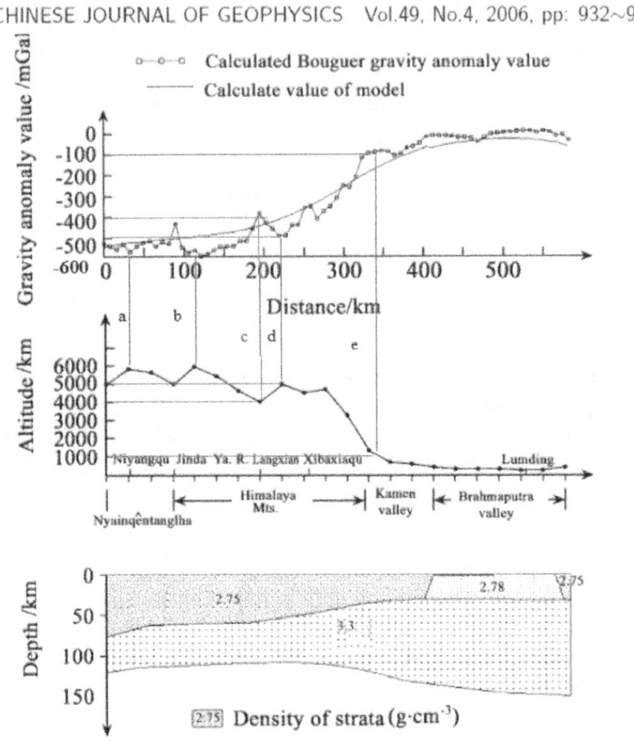

Fig. 7 Crustal structural profile $C - C'$ inversed
from gravity field data of the 'east structural knot'
area and its adjacent

We would like to appeal to anyone who would be able to take weight measurements at high altitude - e.g. anyone who lives in the Rockies in the USA or anywhere in a mountain range and is able to weigh a 100g or 200g weight at sea level and then go to as high an altitude as possible with a sensitive weighing scale[263] and weigh the same weight and compare the results. (This was tried on a holiday flight at approx. 10,000m but the aircraft was simply not stable enough to give a usable reading.) Another possibility might be to use a drone or a balloon, but this would need good stability to work. There is a possibility that if we did measure results that disagreed with the Shell Theorem, it could be put down to local geological factors, so an airborne test might reduce the chances of this affecting the results.

Gravity, Paul Dirac and "Large Numbers," Gyroscopes

As we have already seen earlier, to calculate the acceleration (or force) due to gravity at the Earth's surface, we use this equation:

$$g = \frac{GM}{r_c{}^2}$$

The problem we are trying to address is why this force has seemingly increased since the time of the dinosaurs, some 300 million years ago. Assuming the basic validity of this equation/ relationship, for the value of "g" to have changed, one or more of the other values must have changed.

We have already stated that we don't think the mass of the Earth has changed, although having said this, the amount of mass is often determined by the weight, so this can present some problems when we don't have a way of directly measuring the mass, by for example, putting it on a scale.

It was in the 1940s that physicist Paul Dirac[264] proposed that one of the other figures in our equation G (the so-called gravitational constant) may not be quite so constant.

In a section titled "the variation of the constant of gravitation with time," a posting on Britannica.com reports[265]:

> The 20th-century English physicist P.A.M. Dirac, among others, suggested that the value of the constant of gravitation might be proportional to the age of the universe; other rates of change over time also have been proposed. The rates of change would be extremely small, one part in 10^{11}per year if the age of the universe is taken to be 10^{11} years; [measuring] such a rate is entirely beyond experimental capabilities at present. There is, however, the possibility of looking for the effects of any variation upon the orbit of a celestial body, in particular the Moon. It has been claimed from time to time that such effects may have been detected. As yet, there is no certainty.

As the posting essentially says, it would be probably difficult to measure long term variations over periods of more than about 500 years - because that is about the age of western science - between 5 and 7 human lifetimes!

Another article on Einstein online discusses the matter further and raises another interesting consideration[266]: "… is G really constant, or is it a function of the *matter distribution* in the universe?" The article then goes on to note:

> There is one particularly interesting point of similarity between inertial forces and the gravitational force. An inertial force such as [the] centrifugal force is proportional to the mass of the object being pulled. The same is true for gravity, the gravitational force that pulls an object towards Earth. This force (also called "weight") is also proportional to the mass of the object it acts on.

This is interesting to me because of what I wrote about the findings of Eric Laithwaite and Bruce De Palma in chapter 7 of my book "Finding the Secret Space Programme."[8] We can also consider that the Earth rotates - so does the change in radius affect more than just the centripetal force? Does any change in the radius and rotation speed of such a large object have some less obvious effect - such as the anomalous effects that Eric Laithwaite documented in his demonstrations with gyroscopes[267]?

At this point in time, we can say that there is no agreement (in the white world) as to what causes gravity and "how it works." However, we can say with a high degree of certainty that someone else knows a lot more about how gravity works than we do, because of what happened on 9/11[268]. (This is the subject of two of my other books if you want to learn more[269].)

In the next chapter we will look more at how we might have to re-consider "how gravity works" and what causes it, to help us explain why 300 mya, the force of gravity on the Earth's surface was almost certainly less than its current value..

On the Orbit of the Earth and the Moon…

Both Maxlow and Adams have suggested that our Earth's mass has increased from the creation of new material within our planet. If that were true the mass of the Earth would have increased by a factor of at least 4 – and Stephen Hurrell (as mentioned earlier) suggests an Earth mass increase factor of 8. This has serious consequences for our relationship with the moon. Increasing Earth's mass by a factor of 4 or 8 would, in turn, increase the gravitational pull between ourselves and the moon. If this had truly happened, it would have resulted in a decaying moon orbit and, very likely, the Earth's destruction!

Some researchers, such as Keith Hunter have noted the slowing of Earth's rotation and lengthening of the Earth's orbital period since antiquity[270]. Hunter and others suggest the Earth had a 360-day year compared to our current 365.25 day orbital period.

Keith Hunter also mentions references to a diminished or reduced size of the moon. We suggest that an increased length of year might be due to the moon's loss of mass. As observed by Neal Adams, the moon shows signs of expansion on its surface[271]. If water was vented during this expansion (steam powered expansion) and since gravity (or electric charge – see later) on the moon is insufficient to retain that water, water/mass has been lost by the moon - resulting in less attraction between Earth and Moon. Reduced attraction has led us to "drift apart" hence the moon appears smaller!

The reduced mass of the moon means the combined mass of the Earth and moon has also reduced. The reduced mass of the Earth-moon system would weaken the attraction to the Sun and so perhaps lengthen the orbital period to 365 days.

"It's Only Rocket Science"

During one of NASA's early ventures into space - the Explorer 1 mission[272] (which is regarded as the mission which discovered the Van Allen Zones), something unusual happened. The probe was launched and was tracked by Cape Canaveral and Goldstone (Australia). However, it "arrived" in the expected location about 12 minutes late[273]. This seems to have been because its orbit was higher than was planned - by a considerable margin. One explanation for this is a miscalculation of the burn time for the rocket, or a deviation in its trajectory during launch. However, what if the force of gravity reduced more rapidly than expected with increase in altitude?

Lunar Probe Missions in the 1960s

Wikipedia (for all its faults)[274] has a useful list of lunar probe missions[275]. Inspection of the outcome of these missions would not have given Apollo Astronauts (or rather "Actornauts") much confidence in their own endeavours. For example, all the missions in 1958, 1962 and 1963 failed. Even in 1969, all but one of the USSR missions failed (there were no USA Lunar probes launched in 1969).

We have already considered the possible effect of a different Centre of Mass Force on the Explorer 1 probe. So, whilst one could easily argue that most of these failures were due to the fact that these missions all took place in the "early days" of space exploration, what if a proportion of the failures were due to calculations based on incorrect assumptions about the variation of the force of gravity when going into orbit around the moon? Did scientists eventually learn to add in "fudge factors" to make sure the probes went into the right orbit? In 2006, NASA published an article called "Crash Landing on the moon"[276] which states:

> Crashing was much easier than orbiting, they discovered. The **Moon's uneven gravity field tugs on satellites in strange ways**, and without frequent course corrections, orbiters tend to veer into the ground.

If there are any experts in orbital mechanics out there who have worked on space missions and know of any "fudge factors" that may be applied in calculations, we'd love to hear from them!

Gravitational Binding Energy

One objection to the GPEE theory might be the consideration of Gravitational Binding Energy (GBE). If one considers that energetic processes were in play during the formation of the Earth and the Earth coalesced from smaller pieces, then the force of gravity had to eventually overcome the forces/energies which were present in this collection of pieces. It is generally considered that the coalescing of the Earth created large amounts of heat - as the kinetic energy of pieces of material colliding with each other was converted into heat. The overall energy should then be related to the GBE. The GBE for a sphere[277] is given by

$$U_G = \frac{3}{5}\frac{GM^2}{R}$$

However, if the original Earth's radius was only about half of the present radius, then the gravitational binding energy would also be half the accepted figure. Similar, if we accept Woodhead's argument about greatly reduced mass, the Gravitational Binding energy would also be very much reduced - to about 1/50th of the currently accepted figure.

Further Lunar Considerations

In considering the GPEE theory, let us assume that the moon formed in the same accretion zone as Earth. Imagine the two spinning bodies of similar composition. The moon would, over time, undergo a similar evolution as previously described i.e. heating, expansion and venting, with an emphasis on venting, as the moon influences our seas, and by implication, any subterranean water. The Earth's influence on any subterranean water on the moon would be significantly greater.

Lunar Seismic Activity

Again, this is mentioned because it is possible that the moon, too, is undergoing gas-powered expansion - which could explain some of what are known as Transient Lunar Phenomena[278]. For example, in 1968, a document was published called "NASA Technical Report R-277[279]." Perhaps some of these phenomena were caused by the effects of expansion. Also, Lunar Reconnaissance Orbiter discovered evidence of recent volcanic activity[280].

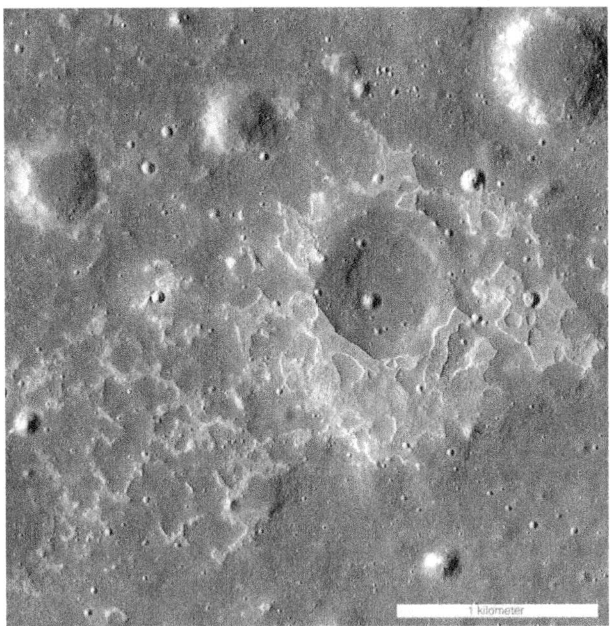

Caption: The feature called Maskelyne is one of many newly discovered young volcanic deposits on the moon. Called irregular mare patches, these areas are thought to be remnants of small basaltic eruptions that occurred much later than the commonly accepted end of lunar volcanism, 1 to 1.5 billion years ago. Image Credit: NASA/GSFC/Arizona State University

An image released as part of the Lunar Reconnaissance Orbiter's "GRAIL" mission[281], with the caption is shown below.

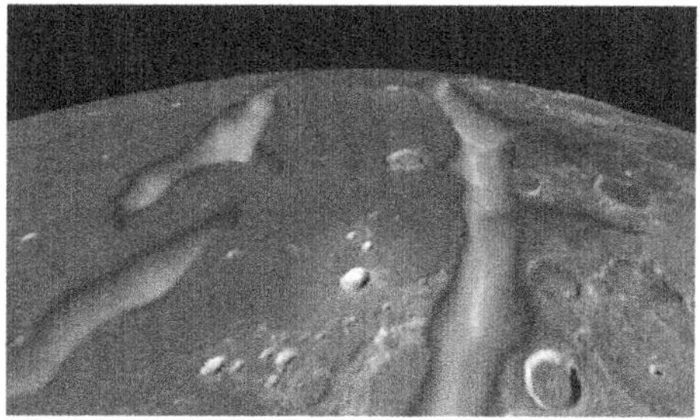

A view of Earth's moon looking south across Oceanus Procellarum, representing how the western border structures may have looked while active. The gravity anomalies along the border structures are interpreted as ancient, solidified, lava-flooded rifts that are now buried beneath the surface of the dark volcanic plains, or maria, on the near side of the moon.

Maria

In October 2014, Space.com reported new discoveries about very large scale lunar geological features in an article entitled "Strikingly Geometric' Shapes Hidden on Moon's Surface[282]." Could these be the result of gas-powered expansion, which on Earth has resulted, ultimately, in the formation of continents? Further, an article published on the "earthSky" Website on 02 Oct 2014 suggests a "New origin for mysterious lunar Ocean of Storms."[283] It states

An ancient asteroid impact was thought to have created moon's Ocean of Storms. Now scientists think it formed via processes within the moon itself.

This is a composite image of the lunar far side taken by the Lunar Reconnaissance Orbiter in June 2009, note the absence of dark areas.

It is true that there are very few maria on the far side of the moon[284]. Could it be that the fluid or material that caused the dark colouration of the maria came from a fluid-bearing layer - like that proposed as being below the mantle? Again, this fluid may have had more of a tendency to erupt or exude on the side of the moon which faces the Earth, due to the stronger tidal forces and difference in shell/crust thickness.

Tidal Locking

The Moon is now in a Synchronous Orbit[285] (its rotation period is equal to its orbital period around the Earth). For tidal locking to occur, the satellite has to be close enough to the host planet[286]. An "Ask an Astronomer" posting about Tidal Locking on the Cornell University website states:

Almost all moons in the Solar System keep one face pointed toward their planet. (The only exception we know of is Hyperion, a moon of Saturn.) This tells us it's probably not a coincidence, that there is probably a reason for this to happen, a physical process that happens to most moons to slow their rotation.

*That process is called **tidal friction**. You probably know that the Moon's gravity affects the Earth's oceans - but then obviously, the Earth's gravity also affects the Moon. It distorts the Moon's shape slightly, squashing it out so that it is elongated along a line that points toward the Earth. We say that the Earth raises "tidal bulges" on the Moon.*

The Earth's gravity pulls on the closest tidal bulge, trying to keep it aligned with Earth. As the Moon turns, feeling the Earth's gravity, this creates friction within the Moon, slowing the Moon's rotation down until its rotation matches its orbital period exactly, a state we call tidal synchronization. In this state, the Moon's tidal bulge is always aligned with Earth, which means that the Moon always keeps one face toward Earth.

Here, we note a very interesting use of the phrase "tidal friction." We are well aware of the moon's tidal effect on our oceans and the effect is demonstrable. We note this phenomenon here, as it seems reasonable to suggest that if there is indeed a layer of water beneath the crust of planets and planetoids, it could help to partly explain why "tidal locking" occurs. It seems more likely to occur if the body had (or has) a higher proportion of liquid under its crust. Additionally, if the amount internal tidal friction might be higher if the liquid is of lower viscosity (i.e. more "runny" or "watery.")

14. The Expanding Earth and the Electric Universe

In this chapter, we will consider the so-called Electric Universe model, proposed by Wallace Thornhill[287] and others. What relevance might it have to the change in the acceleration due to gravity at the Earth's surface?

The Electric Universe

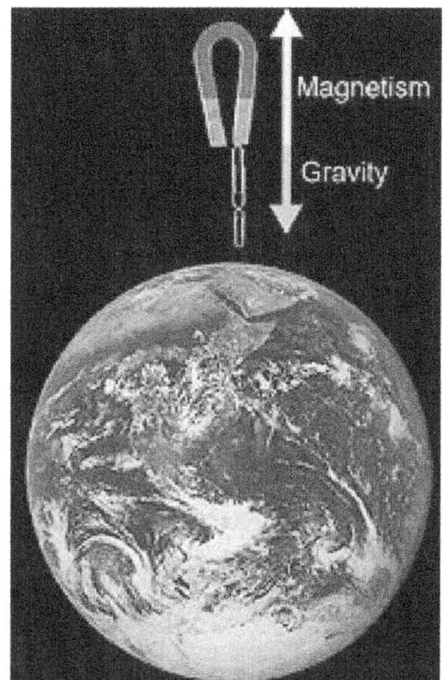

One of the underlying pillars of this way of looking at things is to acknowledge that the electromagnetic force is many trillions of times stronger than the force of gravity - and this is easily demonstrated whenever you use a magnet to pick up a small object - overcoming the gravitational pull of the whole Earth! The electrical force and magnetic force are inter-related.

EU proponents cogently argue that the large-scale structures in the Universe are shaped by electrical phenomena much more than gravitational phenomena.

They argue that in many cases, cosmological forms such as the shapes of galaxies can be explained better in terms of being the result of the flow of electrical currents rather than the result of gravitational forces[288] (see below).

One of the main proponents and developers of the basis of the Electric Universe theory is Wallace Thornhill[289], an Australian Physics graduate who had considered a career in the academic world, but….

> *…he had been inspired by Immanuel Velikovsky through his controversial best-selling book, "Worlds in Collision." Wal experienced, first-hand, the indifference and sometimes hostility toward a radical challenge to mainstream science. He realized there is no career for a heretic in academia.*

So, he went to work for IBM Australia, in developing an early computer graphics system. In this role, he had access to University libraries and a number of scientists and this situation helped him to develop the Electric Universe model further. On his website, Thornhill writes[290]:

> *It is now a century since the Norwegian genius Kristian Birkeland proved that the phenomenal 'northern lights' or aurora borealis is an earthly connection with the electrical Sun. Later, Hannes Alfvén the Swedish Nobel Prize winning physicist, with a background in electrical engineering and experience of the northern lights, drew the solar circuit. It is no coincidence*

Thornhill's website (www.holoscience.com), has a wealth of information and is updated regularly. In 2005 and 2007, Thornhill and self-taught comparative mythologist, David Talbott produced books[291] describing the Electric Universe model and how events described in myths could very well be the result of large-scale plasma discharges. Thornhill is also involved with producing very interesting YouTube videos[292] in response to news items which appear in the fields of Physics, Cosmology and Astronomy.

One example of the fundamentally electric or electrical nature of cosmological phenomena is that of planetary nebulae, such as the beautiful hourglass nebula:

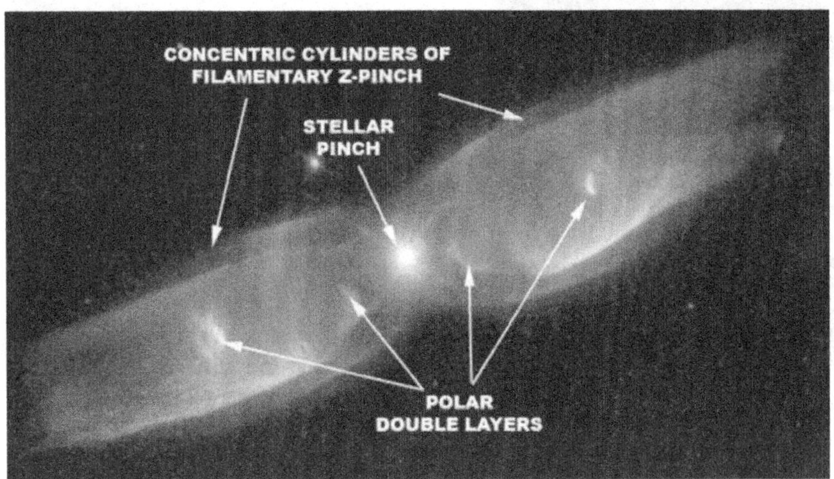

EU Proponents make a good case that such formations are produced as a result of Birkeland Currents[293].

Scientists from many different disciplines contribute to the EU research community and so it is worthwhile investigating the various videos they produce, or visiting one of the yearly EU conferences.[294]

Gravity in an Electric Universe

In 2008, Wal Thornhill posted an article entitled "Electric Gravity in an Electric Universe,"[295] which is of great interest to us, in trying to solve the "Earth expansion/force of gravity" issue. Thornhill considers the problems of "mass" and "weight" and he notes, for example, that Newtonian Gravitation mathematics and explanations do not relate to time. He also makes an interesting observation about comets:

appearance, together with the recently recovered high-temperature minerals (rock particles) from a comet, that give the accurate picture. Comets and asteroids are fragments of planets. They are not primordial - quite the reverse, in fact.

Could this give us a clue as to why the Earth's mass may not be quite what is accepted - which in turn would mean that the force of gravity could have varied? (While on the subject of comets, I can recommend the video "Thunderbolts" has produced on this topic[296].)

Thornhill also considers how force is transmitted to an object to make it accelerate:

*But when we apply force to a body, how is that force transferred to overcome inertia? The answer is 'electrically' by the repulsion between the outer electrons in the atoms closest to the points of contact. **The equivalence of inertial and gravitational mass strongly suggests that the force of gravity is a manifestation of the electric force.***

The reference to inertia is also interesting, and it is something that has to be considered with rotating masses (i.e the Earth!)

Thornhill further goes on to point out the behaviour of protons and electrons when they are accelerated in a strong electrical field. He carefully considers the difference in the accepted mass of the proton and the electron, thus:

*The 2,000-fold difference in mass of the proton and neutron in the nucleus versus the electron means that gravity will maintain charge polarization by offsetting the nucleus within each atom (as shown). **The mass of a body is an electrical variable - just like a proton in a particle accelerator.** Therefore, the so-called gravitational constant... That is why 'G' is so difficult to pin down.*

This difference in mass seems a fairly obvious point that most physicists overlook. That is, if the proton is much "heavier" (more massive) than the electron, surely this must be important in any consideration of the relationship between gravity and electric charge? Was this an idea that physicists such as Paul Dirac ever considered? Developing this idea further, Thornhill says:

Conducting metals will shield electric fields. However, the lack of movement of electrons in response to gravity explains why we cannot shield against gravity by simply standing on a metal sheet. As an electrical engineer wrote, "we [don't] have to worry about gravity affecting the electrons inside the wire leading to our coffee pot." If gravity is an electric dipole force between subatomic particles, it is clear that the force "daisy chains" through matter regardless of whether it is conducting or non-conducting.

To his credit, Thornhill then mentions the work of Dr Eugene Podkletnov, which I discussed in chapter 11 of "Finding the Secret Space Programme."[8] Thornhill reports:

This offers a clue to the reported 'gravity shielding' effects of a spinning, superconducting disk. Electrons in a superconductor exhibit a 'connectedness,' which means that their inertia is increased. Anything that interferes with the ability of the subatomic particles within the spinning disk

> *to align their gravitationally induced dipoles with those of the Earth will exhibit antigravity effects.*

Thornhill then correctly (as far as I have been able to determine) argues:

> *The confusion about any role for electricity in celestial dynamics has come about because of our ignorance of the electrical nature of matter and of gravity.*

Dr Gerald Pollack and the Earth's Magnetic Field

One of the speakers at several EU conferences has been Dr Gerald Pollack who has a Ph.D. in Biomedical Engineering and a B.S.E.E. - Electrical Engineering.[297] He has been studying one of the simplest substances we know of - water - and has found some unexpected things. He notes earlier work by Albert Szent-Gyorgi[298] and Gilbert Ling[299] in relation to the electrical properties of water - although they were considering this in the field of cell biology.

In his EU presentations, Pollack explains his research into how water holds and transmits charge and how, for example, "boundary layers" may form and be exploited for a number of purposes. (Some of his research has been used in applications[300] such as filtration and desalination.)

He has studied the way pH affects charge movement in water and he has also studied the effects that light and sunlight can have on this same charge movement. He, perhaps like Wilhelm Reich, thinks that water can be considered as a kind of "energy sponge" rather than just a liquid. Pollack discusses these matters in his book "Fourth Phase of Water: Beyond Solid, Liquid & Vapor."[301]

As well as considering how water behaves inside cells, Pollack has discussed how water behaves in the Earth's atmosphere and how this is interconnected with the nature of Earth itself. For example, he notes that the atmosphere is (overall) positively charged and that the Earth itself is negatively charged[302].

This can have enormous effects which most scientists don't fully appreciate, even though these charges have been known about and measured for decades.

In his 2015 EU talk, Pollack suggests that the motion of the atmosphere could create the Earth's magnetic field. He includes the following illustration[303]:

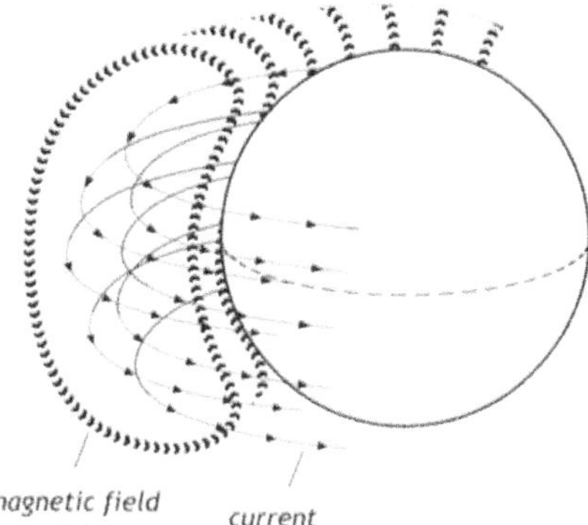

Dr Gerard Pollack's Diagram suggesting how Earth's magnetic field could be created in the atmosphere.

Pollack then suggests:

> *there's charge flow around the Earth… Basically we have a current that's flowing around the Earth. So think of a current now flowing around the Earth from west to east and think about what we know from basic electrostatics and dynamics- if we have a current flow going this way we have a magnetic field … the horizontal arrows - the wind, that's the jet stream that blows constantly around the Earth, it's a current, the current is flowing around the Earth and if you have a current flowing that way then those thicker lines show the magnetic field… just from those charged gradients creating current that flows around the Earth. …We like to think that the Earth's magnetic field comes somewhere from the core of the Earth but you know, nobody's ever been there to make that measurement … so I mean we're really not sure that the magnetic field arises from the core of the Earth and I'm suggesting to you that it's possible that the Earth's magnetic field arises externally and not internally.*

Therefore could the motion of the atmosphere and currents in the Earth's crust combine to create a coherent magnetic field? What other charges or currents might be flowing deeper in the mantle? What similar charge-related effects might occur as a result of a water or steam-filled core?

In relation to Pollack's idea, Peter Woodhead mentioned to me what he thought about reversal of the Earth's magnetic field:

> *If the Earth's magnetic field reverses every 40,000 years or so what is the cause or mechanism? Are we supposed to think Earth's rotation slows down and then reverses? The obvious solution to me is that the Birkeland current between the sun and Earth flips polarity!*

Atmospheric Pressure and the Sun's Energy

In the same lecture, Pollack suggests an alternative cause for atmospheric pressure - which isn't gravity, rather, it's caused by the force of attraction between the negatively charged Earth and the positively charged atmosphere. Pollack suggests that the Sun is causing most of the charge separation:

> *...the Sun is the "ultimate driver" because remember, the sun's energy is what splits the charge in the water which gives rise to the positive charge and so it all comes back to the energy from the Sun. And so the conclusion from this is that the reason the Earth keeps spinning is, it gets energy from the Sun to make it spin. It just doesn't do it. It's known that the rotation speed of the Earth is not constant. It fluctuates even by the day. If the Earth turned by inertia, there shouldn't be fluctuations. There are fluctuations and so this can explain it.*

From the phenomena that he discusses and explains in his 2015 talk, Pollack notes the following facts:

- Charges can exert enormous forces
- Earth bears negative charge; atmosphere positive

Night-time regions have less atmospheric positivity He then suggests an interesting list of possible conclusions:

- Attraction could create atmospheric pressure
- The sun's energy builds atmospheric positivity
- Lateral charge gradients drive winds
- Winds create friction, which may drive the Earth's spin
- Lateral charge flow may create Earth's magnetic field

We will now go on to consider in a bit more detail the possible implications of the positively charged atmosphere and the negatively charged surface of the Earth - to see if that can help us explain the change in the acceleration due to gravity as the Earth has increased in size.

Fredrik Nygaard, The Earth and Gravity

After we had posted the original articles on gas-powered Earth expansion, I was contacted by a chap called Fredrik Nygaard - who completed a BSc in Electrical Engineering in Surrey UK. Fredrik then went into programming, as this was his real area of interest. (At the time he attended University, courses in computer programming of the sort he was interested in weren't available). I interviewed Fredrik on 9th Oct 2016[304] and he has since compiled his research into a free book[305] and a website[306].

Before the development of his own website, Nygaard sent me several articles/essays in which he developed some ideas and calculations on how the Earth's gravity has changed[307] because of what might be loosely-termed "electrical effects." Fredrik Nygaard's work makes similar arguments to those, of Wal Thornhill - and Nygaard also considered gravity might be a "dipole" type of phenomenon, rather than "a monopole." (Some of the quotes I have included below are based on comments Nygaard sent me when he reviewed a draft version of this text.)

In the first section of his book / website Nygaard explains how he thinks Morton Spears' "Particle Quanta" theory is relevant to how these dipoles could relate to the force of gravity. Spears considers the difference between protons and neutrons in a somewhat different way to the "mainstream" quantum theory.

Later, in a section in his book and on his website entitled "Hollow Planets,"[308] Fredrik Nygaard writes:

We know from measuring the electric potential gradient of our atmosphere that our planet is negatively charged relative to the ionosphere. The potential difference is about 300,000 volts. It is the potential difference between the ionosphere and the surface of our planet that keeps our atmosphere from escaping into space. The much weaker gravitational force would not be able to do this on its own.

The negative charge on the surface of our planet is most likely matched with a corresponding positive charge at its centre. This would mean that there is a repelling electrical force inside Earth.

Since gravity is measured from the centre of astronomic bodies, and not from their surfaces, as is the case with the electrostatic force, there can be no net gravity at the centre of planets, moons and stars.

This means that there is nothing to prevent astronomic bodies from being hollow. There is no force at the centre of such bodies to counter the effect of internal electric repulsion. Nor is there anything to counter centrifugal forces due to spin.

If a cavity was to develop inside an astronomic body, there would be no way to make it disappear.

Nygaard then discusses how Newton's Shell Theorem also suggests that cavities could form inside a planet. He then notes (as we did earlier) that Edmund Halley also suggested the Earth may be hollow and that Newton did not disagree at that time.

Nygaard also argues that:

Newton's Shell Theorem is predicated on a uni-polar model of gravity. If gravity is affected by a body's capacitance, or if Thornhill's view is correct, then the shell theorem has to be disregarded, at least for near Earth considerations.

He then suggests:

Gravity variations are, after all, used by geologists in order to detect minerals. If our entire planet has an uneven distribution of matter densities, all near Earth gravity measurements may be skewed. That would mean that we have to go a fair distance up and away from our planet for Newton's simple model to hold.

Nygaard then discusses how most people assume that the strength of the surface gravity means that our planet must be made of something very dense. In reference to the accepted view of the Earth's core, Nygaard writes:

The latest estimate is of a super-dense crystal at Earth's core. This material, which only exists in theory, and no-one has ever been able to produce in a laboratory, has all sorts of fantastic properties. This is all required in order to reconcile observed seismic and gravitational data with current theory.

Earth as a Capacitor

Offering an alternative to the "dense core" assumption, Fredrik Nygaard writes:

However, there is a simple way around this. By recognizing that our planet is a gigantic charged capacitor, we can make the proposition that the dielectric material inside capacitors will add to the gravitational force when sufficiently charged.

Building on something we know to be true - that the surface of the Earth is charged, he then writes:

Most likely, our planet is fully charged, which would make the total charge carried by our planet truly enormous. This charge exerts stress on the crust of our planet. Positive quanta are pulled towards the negative surface. Negative quanta are pulled towards the positive surface. Since both protons and electrons are dielectric, an internal stress develops.

(Fredrik originally developed this idea in an article he wrote in 2016 called "The Dipole Model of Gravity and the Expanding Earth," though his thinking has changed somewhat in relation to if or how the dipole gravity model would work, due to difficulties explaining planetary orbits.)

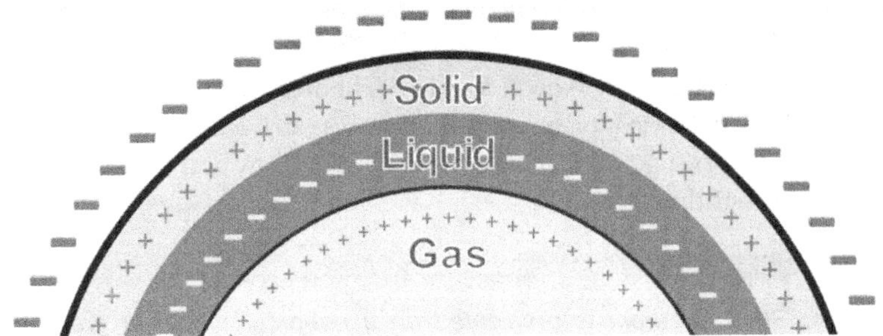

Based on the Nygaard "Earth Capacitor" theory, the suggestion is that if the charge on the Earth's surface is negative, lower layers may end up with alternating/positive negative charges. Also, processes occurring in each layer (material flow, cracking and venting) may affect the charge build up.

Importantly, Nygaard then discusses Gravity Anomalies that have been measured in different places on the Earth's surface (which we discussed in chapter 13) - and how this could be related to geological activity.

The gravitational force is not equally distributed across our planet. Some places have more gravity than others. This is true, even when measurements are adjusted for height above sea level and the centripetal force of our spinning planet. These gravity anomalies are not randomly distributed. They coincide with geological activity. Places with a lot of geological activity have stronger gravity than areas that have little geological activity. It does not matter if the geological activity is due to uplifting of mountains, or formation of rifts.

Nygaard notes, for example that Iceland (situated on the mid-Atlantic rift and geologically very active), has stronger than average gravity[309]. He reports the same is true for the Himalayas and Andes and then mentions areas of lower geological activity also have weaker gravity. Nygaard argues that these changes in local gravity are probably not due to local mass concentrations:

Conventional theory holds that mass alone is the source of the gravitational force. The anomalies are therefore explained by a greater abundance of especially dense matter in the geologically active zones. Dense matter floats up through less dense matter in both regions of rifting and uplifting.

Nygaard then points out that when all other things are equal, the ability to hold electrical charge - i.e. an object's capacitance - increases with surface area. So if the gravity is really due to electrical capacitance then:

...the capacitance of a thin capacitor is greater than the capacitance of a thick capacitor. An expanding hollow planet would therefore be increasing its capacitance, and this would be especially noticeable in areas where the capacitor is cracking.

Bringing these points and ideas together, Nygaard explains:

If the role of capacitance as a source of gravity in our planet is greater than the role of inertial mass, then surface gravity will increase with expansion. The reduction in overall density due to a thinner crust will be made up for by greater capacitance.

An expanding planet will display two types of cracks. There will be rifts where the old crust is pulled apart, and there will be mountains where the old crust breaks in order to fit onto the larger sphere. In both cases, we end up with a thinner crust along the cracks than in areas where there is no cracking. The geologically active areas will have more capacitance, and therefore more gravity than the geologically inactive areas.

It then becomes easier to see his argument regarding the current (small) surface gravity differences because:

- Geologically inactive Tibet and north-east Canada have thick crusts.
- Rift zones like Iceland have thin crusts
- Uplifting cracks like the Himalayas and Andes have thin crusts

I recommend that interested readers do study Frederik's work, as he goes into quite a lot of other relevant detail about his conclusions and he considers how the dipole/capacitance model of gravity could explain both long range and short-range gravitational forces.

Nygaard on Centre of Mass Force

If we consider the capacitance model, Frederik Nygaard suggested that the "Centre of Mass Force" might be experienced in a slightly different way to that proposed by Peter Woodhead. He suggested the following illustration.

A distant observer will have his CMF relatively close to the geometric centre, while an observer close to the surface will have his CMF closer to the surface. See illustration, left.

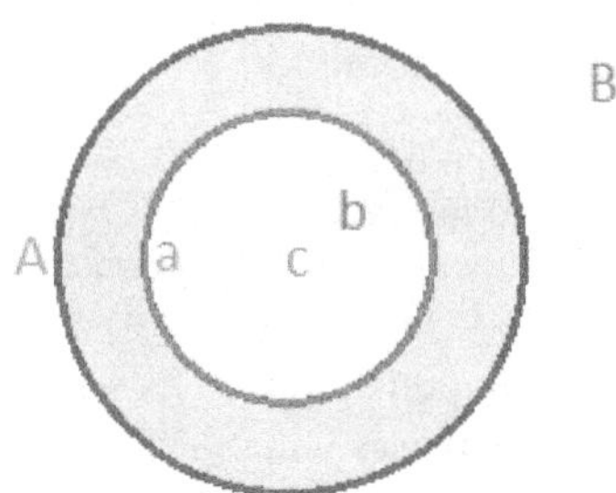

Observer **A**, **B** and **C** are located at different distances from Earth. Their CMFs are located correspondingly at location **a**, **b** and **c**. We can even add an observer **D** inside, next to **a**. His CMF will be located inside the crust.

We seem to have ventured into the same sort of territory that the mathematician Euler did in 1833 – as we discussed in chapter 2!

Coulomb's Law and Universal Gravitation

Those who have studied physics to a higher level may remember Coulomb's law - relating to electric charges. It is worth mentioning here how analogous it is to the law of gravitation,[310] which is expressed thus:

$$F = G\frac{m_1 m_2}{r^2}$$

Where m_1 and m_2 are the masses involved and r is the distance between them. G is the gravitational constant.

Coulomb's law is expressed thus

$$F = k\frac{q_1 q_2}{r^2}$$

Where q_1 and q_2 are the charges involved and r is the distance between them. k is Coulomb's law constant. Perhaps this is more important than most physicists have previously considered, even though most would just say that charges and masses behave as "point sources" in both these equations and they wouldn't think any more about it. Also of note is that with electric charges, the force between them can be repulsive - which, according to current "white world" physics, can never happen with gravity/masses.

Halton Arp and Mass Condensation

Another interesting consideration that Fredrik Nygaard writes about is Halton Arp's theory of mass condensation[311]. Halton Arp (1927-2013), dubbed by some as "the world's most controversial astronomer"[312] was an American astrophysicist who worked at the Max Planck Institute for Astrophysics. Arp carefully studied hundreds of radio telescope and other astronomical observations and in them, he noted that objects that were adjacent and seemed to be structurally associated with one another had different "red shifts." For example, what some claimed were distant quasars were actually connected to nearer galaxies - according to Arp's observations - yet had very different red shift values. Arp went further and suggested that the quasars had been ejected from the galactic centre. The "red shift" here is meant, according to mainstream astrophysics, to indicate how far away - and how fast - an object is travelling.[313] The theory of what causes "red shift" is fundamental to the current "Big Bang" and an "ever-expanding universe" model which has been predominant in astronomy and cosmology since the early twentieth century. (It is the same sort of "settled" model as the fixed-radius Earth model!)

Arp disagreed with the "Big Bang" model, which was further developed using Einstein's relativity theories. However, Arp and others have, again, pointed out that the "Big Bang" and "Inflation" theories cannot explain the many anomalous red shift observations that have been made. He therefore suggested a completely different theory, which included the concept of "matter condensation" over time. Arp argued that matter becomes more dense with age - and the further it is away from the galactic centre. As Fredrik Nygaard explained[314] to me:

> *Halton Arp suggested that radiation in the form of high energy photons condense onto matter, thereby increasing the mass of matter over time while simultaneously cooling down the environment.*

Arp also argued that the changes in density were "stepped" - depending on how far away the matter had moved from the galactic centre. This, he suggested would explain the so-called "quantized red shifts." This latter phenomenon is accepted - that with the "Big Bang," there should be a continuous range of distances of stellar objects and therefore a continuous range of "red shift" values. However, the red shift values that have been observed/calculated fit into about 6 different groups.[315]

Hence, if the matter the Earth is made from has been aging since its formation, maybe the measured density/mass *has* increased in the manner Arp suggests - and the force of gravity, experienced at the surface of the Earth has also, therefore, increased.

Jupiter's "Large and Diffuse Core"

In June 2017, Fredrik Nygaard alerted me[316] to a new posting from NASA - about the Juno mission which has been in progress around Jupiter since July 2016[317]. The posting, with the title shown above[318], in one paragraph reports:

> *Other measurements from Juno hint that Jupiter has a surprisingly large core made of heavy elements. Each time Juno whizzes past the planet - just a few thousand kilometres above the cloud tops - Jupiter's gravitational tug nudges the spacecraft's orbit. Team scientists have analysed that nudge and calculated that the planet has a core that amounts to some 7–25 times the mass of the Earth. **The core could be both larger and more diffuse than**

> **expected**, *extending out to as much as half of Jupiter's 70,000-kilometre radius.*

Is the article trying to suggest that a "diffuse" core might not be a core - i.e. that it might be a… void…?

This article came out several months after Fredrik Nygaard had produced his initial essays and articles in which he developed his ideas, so it was interesting to see what Wallace Thornhill had to say about this new NASA discovery. On 5 July 2017, a video was posted on the "Thunderbolts Project" YouTube Channel[319]. In the video, Thornhill goes through some of the statements made in a NASA news release about the new discoveries from the Juno probe and at the 6:55 mark in the video, Thornhill states:

> *So the inside of Jupiter is unlike anything ever contemplated by Newtonian physics. Jupiter's gravity is almost 2.5 times stronger than that of the Earth. So internally at some depth there will be a strong repulsive force due to the positive gravitational poles facing inwards, which can hold a spherical shell of matter stably in place. Jupiter is hollow and doesn't have a core of liquid metallic hydrogen. Of course, the same argument for hollowness holds for the Earth and all other celestial bodies, so it is significant that a number of leading scholars in the past have considered this possibility.*

It seems Thornhill has paid some attention to Frederik Nygaard's research. In the video, soon after the statements he made above, Thornhill, includes some quotes from Jan Lamprecht's "Hollow Planets" book, which we talked about in chapter 4 - including the seismology arguments Lamprecht made. It is worth remembering, however, some of the problems with Lamprecht's research - his book is only a "feasibility study," not a research paper and Lamprecht is only "in the same boat" as the rest of us (i.e. an "interested researcher" rather than someone who has a lot of experience working in a specific area.) Still, Lamprecht's observations and models are important, I feel.

Nygaard on the Formation of the Solar System and Planets

When I read Fredrik's earlier "The Gravity Mystery" article[320], I was particularly impressed with the way he described a theory about the creation of our Solar System and the planets. This is consistent with the ideas included in the Electric Universe and the earlier points we raised regarding the formation of the Earth. Fredrik wrote:

> *Our solar system was most probably created in a supernova event some 4000 million years ago, and the creation process itself took no more than a few days to complete. There was a short circuit in the cosmic current flowing in our region of our galaxy, and matter at near absolute zero temperatures were pulled violently into a sun in the middle, several planets, and a large number of moons.*
>
> *Every planet and every moon got its crust thoroughly roasted. However, the heat was not strong enough to penetrate very deep into these bodies, so all but the sun in the middle quickly cooled down over the next few years.*

But the glowing sun, powered by the same electric current that created it in the first place, kept the planets from cooling down completely. Over millions of years, every planet warmed to the glow of the sun, both directly through radiation, and indirectly through interactions of magnetic fields and auroras.

The internals of the various planets went from ice to liquid and then on to gas. The larger planets, having enormous reservoirs of gas and water trapped inside of them soon cracked and became gas giants. Smaller planets like our own, with relatively much less gas and water trapped in them, did not crack for over a billion years.

15. Considerations and Conclusions

In this volume, I have tried to analyse a wide range of evidence that seems to conclusively show that the creation, development and structure of our planet is quite different from what mainstream sources in academia and the media would have us believe.

When I first started looking into the "Hollow Earth" stories and myths, I was rather sceptical of the validity of the accounts and I could find no credible evidence to support the idea that people have travelled deep into the inner Earth. Neither could I find any evidence of large polar holes or other entrances to this "inner world."

Nevertheless, I then ended up in a "new position" thanks primarily to the research and presentations of Neal Adams, Dr James Maxlow and Peter Woodhead. My new thinking (and deeper understanding in some areas) was further influenced by Wallace Thornhill's Electric Universe research, along with Fredrik Nygaard's (what I consider to be) refinements of the EU concepts as they may relate to the formation of the Earth, in particular, and other planets. I think that together, Peter Woodhead's and Frederik Nygaard's research and thinking could bring us closer to understanding or even proving what is inside our planet and how it formed.

Polar Cosmic Ray Anomalies

Just occasionally, after one has spent many hours compiling and then posting research, hoping to have found all the typos and minor foibles, one is rewarded with some interesting and intelligent feedback, instead of irritating comments or anonymous insults. This was what happened when I recently noticed remarks by David Brelin on my YouTube video "Explaining the Expanding Earth (Part 1) - With Peter Woodhead"[183]. Brelin had *actually watched* the video and in January 2019, had this to say:

> *Concerning the recently observed "inexplicable" cosmic rays emanating from the continent of Antarctica, a water/steam filled Earth core would explain this phenomenon.*
>
> *It should also be noted that since the Arctic is covered by oceanic liquid water - a strong neutron moderator - which would preclude it from radiating the same cosmic ray energy as observed at Antarctica due to the neutron moderating effects of liquid water.*
>
> *If the Earth's core is filled with an oblate spheroid of super-heated gaseous water steam, then that geometry would tend to focus the cosmic ray energy into a divergent beam pattern as it leaves the planet through a kinetic aperture in the absence of liquid water as a neutron moderating force.*
>
> *Since geostationary planetary orbit of satellites cannot be achieved over the polar regions, this precludes satellite observations of such events. Such events are thus constrained to terrestrial observation only in general, and to the southern polar region in particular.*

I checked what he was describing (I had never heard this particular science news item) and it described in an article entitled "Antarctica's Spooky Cosmic Rays Might Shatter Physics As We Know It"[321] (posted on 4 Oct 2018).

Africa is Splitting to Form A New Ocean

Making a similarly useful contribution, someone recently sent me a short video clip which again seems to include evidence of ongoing Earth expansion. In 2006, Geologists witnessed the formation of a rift in East Africa and this was reported in the German "Spiegel" newspaper[322]:

Geologist Dereje Ayalew and his colleagues from Addis Ababa University were amazed - and frightened. They had only just stepped out of their helicopter onto the desert plains of central Ethiopia when the ground began to shake under their feet. The pilot shouted for the scientists to get back to the helicopter. And then it happened: the Earth split open. Crevices began racing toward the researchers like a zipper opening up. After a few seconds, the ground stopped moving, and after they had recovered from their shock, Ayalew and his colleagues realized they had just witnessed history. For the first time ever, human beings were able to witness the first stages in the birth of an ocean.

On the UK Geological Society Website, we can read[323]:

The Arabian Plate is rifting away from the African plate along an active divergent ridge system, to form the Red Sea and Gulf of Aden. The rifting then extends southwards where the African Plate is itself becoming stretched along the line of the East African Rift Valley and is splitting to form two new plates; the Nubian and Somalian Plates.

An interesting 2014 lecture by David Hilton[324] - a geochemist at the Scripps Institution of Oceanography gives further details about how the lithospheric plate is rifting. Again, I would suggest that this is strong evidence of ongoing Earth expansion.

Understanding the Gravity of the Situation...

In researching Earth expansion and paleogravity, we again get into the areas of how "white world" or mainstream physics cannot explain gravity (and therefore cannot engineer the control of it). As I have already said earlier in this volume, in my previous books, I have shown evidence that there is a "black world" and people working within it know how to affect or control gravity. This was proved by what happened on 9/11 (you really should research this properly if you haven't already). I am therefore given to wonder if one of the reasons that Earth expansion (and related) research is not supported well - and largely remains outside the mainstream - is because a detailed study of it would reveal that gravity is, indeed, some kind of electrical phenomenon, as the EU model implies. This area of research is "taboo" and no scientists in the white world have an easy time doing experiments like those of Eugene Podkletnov and Ning Lee (please do your own research and/or read my other books!)

An Expanding Earth, but a Hollow Earth?

I have no doubt that the Earth is undergoing expansion and this is ignored by almost everyone, despite the considerable amount of evidence brought forward primarily by Dr James Maxlow. I would hope that all geologists and scientists will, after more careful study, accept his conclusions, rather than behaving in the manner described by Leo Tolstoy…

Some would question that we could have a hollow or gas-filled core, but we might consider a slightly different arrangement, where we do have a large void in the interior, but not as big as Peter and I suggested. For example, there could be a much larger amount of water than the "standard" core model suggests, but not as much as Woodhead originally suggested - because according to the rough calculations we did, there could still be an enormous amount of water under the mantle - enough to fill the current oceans several times over.

So I conclude that the boiling of water from a formerly frozen core explains the driving force behind the expansion, as Woodhead proposed, but perhaps surface gravity is mostly created by the Earth's electrical capacitance, as Fredrik Nygaard has proposed. Both the capacitance of the Earth and interactions with the electric field in the sun and in the solar system have changed over millions of years, which has caused an associated change in the surface gravity.

With these concepts and ideas now duly laid out and described in detail, I hereby state the following general conclusions:

1. The expansion of our Earth and probably other planets and moons is caused by the heating of their initially frozen interiors and consequent gas-powered expansion.
2. The 4.2 billion years of zero Earth expansion is due to the core remaining frozen.
3. There cannot be any large size polar (or other) openings, as this would not allow enough pressure build up within the Earth.
4. The Earth's surface area, radius and volume are still increasing exponentially.
5. The Earth's interior was, and is, being warmed from its original frozen state by a combination of means. At least 50% of the heat is derived from the nuclear decay of uranium and thorium.
6. The mass of the Earth has not significantly increased in 4.5 billion years.
7. No effects have been observed that might be produced by any increase in mass.
8. Earth's gravity at its surface is increasing exponentially.
9. The ability of giant flora and fauna to exist was very much related to the lower value of gravity at the surface of the Earth, before it expanded.
10. Some species extinctions are likely to have been due to the increasing

force of gravity on plant and animal life.

11. A great outflowing of water might explain the historical "great flood" as described in many cultures - and this explains the creation of the deep oceans.

12. The abiotic production of oil is due to the presence of methane in the core being forced under pressure through the mantle into low regions in the crust. The hot, pressurised water present in the transitional zone is then an ideal region in which chemical reactions - forming all kinds of organic compounds - can take place. This process is happening all the time - as the Earth is still expanding.

13. Gravitational measurements taken by various studies are not anomalous (as always stated), but caused either by an offset CMF, within a gas filled expanded Earth or the fact that crustal thinning affects the local capacitance. That is, the reduction in the acceleration due to gravity with increasing altitude is also better predicted by the GPEE model than the accepted "iron core" model.

14. Despite expansion, Earth's relation to our Sun and Moon have not been substantially affected.

15. The increasing distance between Earth and moon is due to venting of gas/water through the moon's shell creating a loss of mass. The moon has an insufficient force of gravity (or an insufficient amount of electrical charge) to retain an atmosphere - so it has lost mass, while the Earth's mass has not changed much.

16. The Cassini probe identified water under the surface of Saturn's moons Titan and Enceladus. Jupiter's moon Europa has a subterranean ocean, as might Ganymede and Calisto. Neptune's moon Triton is also suspected to have water. These observations support our assertion of there being substantial amounts of water under the Earth's crust and mantle.

Future Earth Expansion

So what of the future? Maxlow predicts continued exponential growth - leading to Earth becoming a gas giant! I have to agree. Earlier, we have calculated that there is still potentially 99% of our estimated original amount of water still under the mantle. If the heating continues, which it will - as there are still radioactive elements in the crust, and if the Earth is indeed heated by telluric currents, there is no reason why the water should not continue to boil.

The continued expansion would cause an increase in capacitance and therefore may cause an increase in the force of gravity at the surface - resulting in smaller sized mammals, smaller sized humans and so on.

Will the pressure be vented bringing an end to expansion? Will we be living on water-world and evolve back into sea dwellers? Perhaps dolphins are even smarter than we think!

If venting of water increases ahead of expansion, then we would predict rising sea levels - not because the Polar Ice caps are melting, but because more water is being forced out from under the mantle due to the gas expansion. We might therefore be given to wonder will all the material in the mantle, ultimately, be "extruded" to form ocean floor? Will all the water be forced to the surface to form oceans with four times their present volume? Or, will the oceans' depths remain constant - or even go down - as the Earth itself continues to expand, and the water is therefore spread over a larger surface area? Will the crust eventually become so thin that it is no longer viable, and the planet eventually explodes? At this point, one is reminded of Tom Van Flandern's Exploded Planet Hypothesis…[325]

Obviously, much more work needs to be done to resolve some of the outstanding questions. It is now up to mainstream science to "catch up" with and accept the findings and evidence compiled by Carey, Maxlow and others which proves the case for Earth expansion beyond reasonable doubt.

Perhaps we can even say that our Earth is a living, growing entity as some have surmised.

Thank You!

In closing then, I would like to thank you for working your way through to the end, and for considering what I have presented to you. I hope you have found it enlightening and useful and that the information herein contributes, in some small way, to a better understanding and appreciation of the world we live on.

16. References

You can find the links used here in an online PDF version of the book at
http://www.checktheevidence.com/ or via this short link:
http://bit.ly/ebnkbook

[1] https://pixabay.com/users/comfreak-51581/

[2] https://pixabay.com/illustrations/lava-cracked-background-fire-656827/

[3] https://www.dailymotion.com/video/x201eo8

[4] https://www.britannica.com/biography/Eratosthenes

[5] https://www.britannica.com/place/Earth/Development-of-earths-structure-and-composition

[6] https://www.britannica.com/science/carbonaceous-chondrite

[7] http://www.psrd.hawaii.edu/Oct04/SaU169.html

[8] https://www.checktheevidence.com/wordpress/2018/05/31/book-finding-the-secret-space-programme/

[9] https://www.natgeokids.com/uk/discover/geography/physical-geography/structure-of-the-Earth/

[10] https://www.britannica.com/place/Earth/The-interior

[11] https://www.smithsonianmag.com/science-nature/when-continental-drift-was-considered-pseudoscience-90353214/

[12] https://www.britannica.com/science/plate-tectonics/Development-of-tectonic-theory

[13] https://geology.com/pangea.htm

[14] https://www.imdb.com/title/tt0074157/

[15] https://www.gutenberg.org/ebooks/18857

[16] https://www.gutenberg.org/ebooks/123

[17] https://www.youtube.com/watch?v=ubKFmIDBMjU

[18] https://www.crystalinks.com/hollowearth.html

[19] https://greekgodsandgoddesses.net/gods/hades/

[20] https://norse-mythology.org/cosmology/the-nine-worlds/svartalfheim/

[21] http://jewishencyclopedia.com/articles/13563-sheol

[22] http://www.kabbalah.info/eng/content/view/frame/2872?/eng/content/view/full/2872&main

[23] https://www.ancient-code.com/shambhala-the-hollow-Earth-according-to-ancient-buddhism/

[24] http://www.bbc.co.uk/history/ancient/cultures/shangri_la_01.shtml

[25] http://www.dioi.org/kn/halleyhollow.htm

[26] https://www.jstor.org/stable/pdf/101831.pdf

[27] http://www.unmuseum.org/hollow.htm

[28] http://eulerarchive.maa.org/hedi/HEDI-2007-04.pdf

[29] https://archive.org/details/letterseulerond03brewgoog/page/n187

[30] https://archive.org/details/letterseulerond03brewgoog/page/n189

[31] https://www.britannica.com/biography/John-Leslie-Scottish-physicist-and-mathematician

[32] https://archive.org/details/elementsofnatura00leslrich/page/450

[33] https://babel.hathitrust.org/cgi/pt?id=wu.89062341235&view=1up&seq=120

[34] https://www.unbelievable-facts.com/wp-content/uploads/2018/03/John-Quincy-Adams-Symmes-Circular-No.-1-1818.jpg

[35] https://www.atlasobscura.com/places/hollow-Earth-monument

[36] http://www.hilobrow.com/2011/09/11/hollow-Earth/

[37] https://babel.hathitrust.org/cgi/pt?id=loc.ark:/13960/t17m1g52t&view=2up&seq=6

[38] https://archive.org/details/hollowglobeorwo00lyongoog/page/n8

[39] https://www.amazon.ca/Agharta-Robert-Dickoff/dp/0787312398

[40] http://www.wikibin.org/articles/robert-ernest-dickhoff-5.html

[41] https://www.bibliotecapleyades.net/tierra_hueca/tierrahueca/contents.htm

[42] https://web.archive.org/web/20051125003727/http:/www.hollowearththeory.com/articles/hollowearthHistory.asp

[43] https://the-eye.eu/public/Radio/Coast to Coast AM/Coast to Coast - 2005/Coast to Coast - Nov 24 2005 - Hour 2.mp3

[44] https://www.coasttocoastam.com/shows/2005/11/24

[45] https://www.britannica.com/place/Japan-Trench

[46] http://tinyurl.com/apuacc

[47] http://checktheevidencecom.ipage.com/checktheevidence.com/pdf/The Smoky God.pdf

[48] https://www.dailymail.co.uk/news/article-3472768/Incredible-moment-sailors-witness-birth-island-Underwater-volcanic-eruption-creates-new-landmass-stunned-yachtsmen.html

[49] https://www.amazon.co.uk/Whalehunters-Dundee-Whalers-Malcolm-Archibald/dp/1841830658

[50] https://web.archive.org/web/20120318230147/http:/www.ourhollowearth.com/PolarOpn.htm

[51] http://www.esa.int/esaCP/SEM7ZF8LURE_index_1.html

[52] http://moteprime.org/article.php?id=6

[53] https://visibleearth.nasa.gov/view.php?id=55099

[54] https://web.archive.org/web/20110716191559/http:/www.ourhollowearth.com/ExpeditionUpdate15.htm

[55] https://www.amazon.co.uk/Evolution-Through-Contact-Becoming-Citizen/dp/1470063956

[56] https://www.youtube.com/watch?v=ZUeGTTzZk-k

[57] https://www.palinstravels.co.uk/book-634

[58] https://www.lonelyplanet.com/attractions/amundsen-scott-south-pole-station/a/poi-sig/363463/1007062

[59] http://palinstravels.co.uk/book-757

[60] https://www.dailymotion.com/video/x7iwe39

[61] https://www.southpolestation.com/trivia/history/dome/dome1.html

[62] http://hollowplanets.com/index.php/shop/

[63] https://www.bibliotecapleyades.net/tierra_hueca/esp_tierra_hueca_8a.htm

[64] https://exemplore.com/ufos-aliens/Hollow-Earth-Theory-and-Admiral-Byrds-Flight-to-Agartha

[65] https://www.britannica.com/biography/Richard-E-Byrd

[66] http://www.phoenixmasonry.org/masonicmuseum/richard_byrd_fdc.htm

[67] https://www.youtube.com/watch?v=w-RLncjmln8

[68] https://www.imdb.com/title/tt0040767/plotsummary

[69] http://www.educatinghumanity.com/2011/03/james-forrestal-early-ufo-facts-and.html

[70] https://youtu.be/w-RLncjmln8?t=2110

[71] https://www.tandfonline.com/doi/pdf/10.1111/j.0033-0124.1956.83_13.x

[72] https://fsa.fandom.com/wiki/Oazisdar

[73] https://www.atlasobscura.com/places/mt-erebus

[74] https://www.bibliotecapleyades.net/tierra_hueca/esp_tierra_hueca_20.htm

[75] http://thehollowearthinsider.com/?s=byrd

[76] https://www.bodymindhealing.info/byrd.php

[77] https://www.coasttocoastam.com/show/2012/06/13

[78] https://web.archive.org/web/20110318012202/http:/www.npiee.org/team.html

[79] https://web.archive.org/web/20110318010824/http:/www.npiee.org/science.html

[80] https://www.sciencephoto.com/media/161061/view

[81] https://web.archive.org/web/20110318010824/http:/www.npiee.org/expedition.html

[82] https://web.archive.org/web/20130303005039/http:/www.indiegogo.com/NPIEE

[83] https://englishrussia.com/2009/07/22/the-kola-superdeep-borehole/

[84] https://www.youtube.com/watch?v=zz6v6OfoQvs

[85] https://www.damninteresting.com/the-deepest-hole/

[86] http://www.crystalinks.com/hollowearth.html

[87] https://www.nature.com/articles/311555a0

[88] https://www.bibliotecapleyades.net/tierra_hueca/esp_tierra_hueca_9.htm

[89] https://www.britannica.com/science/earthquake-geology/Properties-of-seismic-waves

[90] https://www.britannica.com/science/earthquake-geology/Shallow-intermediate-and-deep-foci

[91] https://www.elsevier.com/books/modern-global-seismology/lay/978-0-12-732870-6

[92] https://earthquake.usgs.gov/learn/topics/determining_depth.php

[93] https://plato.stanford.edu/entries/creationism/

[94] https://creation.com/creationism

[95] https://www.youtube.com/watch?v=Jeuv4VSez64

[96] https://www.checktheevidence.com/wordpress/2019/03/08/book-acknowledged-a-perspective-on-the-matters-of-ufos-aliens-and-crop-circles/

[97] https://www.britannica.com/topic/Hinduism/The-Rigveda

[98] http://www.gnosis.org/naghamm/origin.html

[99] https://www.biography.com/writer/ralph-waldo-emerson

[100] https://peoplepill.com/people/antonio-snider-pellegrini/

[101] https://www.age-of-the-sage.org/tectonic_plates/boundaries_boundary_types.html

[102] https://www.Earth-prints.org/bitstream/2122/2017/1/MANTOVANI.pdf

[103] https://www.britannica.com/biography/Alfred-Wegener

[104] https://books.google.co.uk/books?id=l_0l0KOdHLoC&lpg=PP1&pg=PA89&hl=en

[105] https://www.encyclopedia.com/science/dictionaries-thesauruses-pictures-and-press-releases/carey-samuel-warren

[106] https://www.amazon.com/Theories-Earth-Universe-History-Sciences/dp/0804713642/ref=sr_1_10/103-9463814-0048642?ie=UTF8&s=books&qid=1194415368&sr=1-10

[107] https://books.google.co.uk/books?id=2U_gBAAAQBAJ&pg=PA10&lpg=PA10

[108] https://core.ac.uk/download/pdf/33322868.pdf

[109] https://www.nndb.com/people/598/000206977/

[110] http://expansion.geologist-1011.net/

[111] https://www.Earth-prints.org/bitstream/2122/2015/1/Hilgenberg%282%29.pdf

[112] https://archive.org/details/Hilgenberg1933/

[113] https://www.dinox.org/eehistory.html

[114] https://www.dinox.org/sugbooks.html

[115] https://books.google.co.uk/books?id=bfl5gPwcMsIC&pg=PA27&lpg=PA27

[116] https://www.dinox.org/owenpalmap.html

[117] https://www.dinox.org/publications/Owen2018-Expanding Earth Diagrams.pdf

[118] http://www.jamesmaxlow.com/klaus-vogel/

[119] http://www.jamesmaxlow.com/james-maxlow/

120 http://espace.library.curtin.edu.au/R/9NTSHP5DDYRXLE4GXER6QGJ2LGG6AVGQU4B38B965U3466D1A6-00703

121 https://www.amazon.co.uk/s?i=stripbooks&rh=p_27%3AJames+Maxlow&s=relevancerank&text=James+Maxlow&ref=dp_byline_sr_book_1

122 https://www.amazon.co.uk/Terra-non-Firma-Earth-Tectonics-ebook/dp/B00UCL6NR6/

123 https://www.abebooks.co.uk/servlet/SearchResults?isbn=9788825518900&cm_sp=mbc-_-ISBN-_-all

124 https://www.expansiontectonics.com/wpPDF/ExpansionTectonicsHandout0915.pdf

125 https://www.youtube.com/playlist?list=PLFF431CCB43847CBE

126 http://nealadams.com/

127 http://nealadams.com/science-videos/

128 https://www.youtube.com/watch?list=PL5A1097F4E958728D&v=oJfBSc6e7QQ

129 https://www.youtube.com/watch?v=5ZC22-IU7qI

130 https://web.archive.org/web/20031209084412/http:/nealadams.com/earthProject/toon1.html

131 https://web.archive.org/web/20061128053252/http:/www.nealadams.com/earthProject/gravity_and_pressure.pdf

132 https://academic.oup.com/gji/article/13/1-3/349/920314

133 http://www.ngdc.noaa.gov/mgg/ocean_age/data/2008/image/age_oceanic_lith.jpg

134 https://web.archive.org/web/20070305193742/http:/www.platetectonics.com/book/page_12.asp

135 https://www.researchgate.net/publication/23843997_Measurements_of_strain_at_plate_boundaries_using_space_based_geodetic_techniques

136 https://www.nature.com/news/2002/020729/full/news020729-9.html

137 https://pubs.geoscienceworld.org/books/book/430/chapter/3798676/Pangean-shelf-carbonates-Controls-and

138 https://paleobiodb.org/

139 https://www.checktheevidence.com/wordpress/2017/09/24/book-climate-change-and-global-warming-exposed-hidden-evidence-disguised-plans/

140 http://www.pnas.org/content/99/17/10976.full?maxtoshow=&HITS=10&hits=10&RESULTFORMAT=&fulltext=genesis+of+hydrocarbons+and+the+origin+of+petroleum&searchid=1085470440708_510&stored_search=&FIRSTINDEX=0

141 https://www.youtube.com/watch?v=2cUg3lDgJ20

142 https://pubs.acs.org/doi/abs/10.1021/ed031p326

143 http://www.geologie.ens.fr/spiplabocnrs/IMG/jpg/carte_geol_monde-01.jpg

144 https://www.dictionary.com/browse/obduction

145 https://www.dinox.org/expandingearth.html

146 https://www.amazon.co.uk/Dinosaurs-Expanding-Earth-Stephen-Hurrell/dp/0952260379

147 https://www.Earth-prints.org/bitstream/2122/8838/1/Hurrell2011-Ancient Lifes Gravity and its Implications for the Expanding Earth - Erice.pdf

148 https://www.dinox.org/sizelimit.html

149 https://www.dinox.org/publications/Hurrell2019d.pdf

150 https://biblio.co.uk/9780671619466

151 https://www.dinosaurtheory.com/big_dinosaur.html

152 https://www.dinosaurtheory.com/flight.html

153 https://www.britannica.com/science/Cretaceous-Period/Terrestrial-life

154 http://www.krugerpark.co.za/africa_kori_bustard.html

155 http://www.prehistoric-wildlife.com/species/q/quetzalcoatlus.html

156 http://www.prehistoric-wildlife.com/species/a/argentavis.html

157 https://prehistoricparkip.fandom.com/wiki/Arthroplerua

158 https://prehistoricparkip.fandom.com/wiki/Giant_Dragonfly

159 https://www.smithsonianmag.com/science-nature/the-history-of-air-21082166/

160
https://www.researchgate.net/publication/227679835_The_controlling_factors_limiting_maximum_body_size_of_insects

161 https://www.dinosaurtheory.com/solution.html

162 https://www.pgi.gov.pl/en/docman-tree/publikacje-2/special-papers/2340-9-polish-geological-institute-special-papers-9-cwojdzinski-tectonic-structure-of-the-continental-lithosphere-expanding-Earth-pdf/file.html

163 https://archive.org/stream/tolstoyonart00tolsuoft/tolstoyonart00tolsuoft_djvu.txt

164 http://www.jamesmaxlow.com/expansion-tectonics-2/

165 http://www.frontier-knowledge.com/Earth/papers/2011.Eichler.A_new_mechanism_for_matter_increase_within_the_Earth.pdf

166 http://wiki.naturalphilosophy.org/index.php?title=Gerhard_Otto_Wilhelm_Kremp

167 https://www.worldcat.org/title/new-concepts-in-global-tectonics/oclc/24504670

168 https://www.lenr-forum.com/forum/thread/1649-lenr-reactor-based-on-reactions-in-Earth-and-planets-core-updated/

169 https://www.lenr-canr.org/acrobat/Kirkinskiifusionreac.pdf

170 https://web.archive.org/web/20070518141946/http:/www.nealadams.com/PhysicsOfGrow.html

171 https://www.britannica.com/science/pair-production

172 https://www.britannica.com/science/Hawking-radiation

173 http://physics.usask.ca/~bzulkosk/Lab_Manuals/EP353/Absorption-of-Gamma-Rays-BZ-20151123.pdf

174 https://www.youtube.com/watch?v=jWUbj-VRwPk

175 https://www.youtube.com/watch?v=oAN0asAv62Q

176 http://nuclearplanet.com/

177 https://apod.nasa.gov/apod/ap000921.html

178 http://www.nuclearplanet.com/Herndon's Geodynamics.html

179 http://www.nuclearplanet.com/Herndon NCGT 130601.pdf

180 https://www.richplanet.net/richp_genre.php?ref=63&part=1&gen=8

181 https://www.ancient-world-mysteries.com/

182 https://www.richplanet.net/

183 https://www.youtube.com/watch?v=swCnPOi5qOU

184 https://youtu.be/9iIWYYNkgJQ

185 https://www.checktheevidence.com/wordpress/2014/06/12/a-mechanism-for-Earth-expansion/

186 https://www.checktheevidence.com/wordpress/2015/04/14/gas-powered-planetary-expansion-evidence-and-calculations/

187 http://cse.ssl.berkeley.edu/lessons/indiv/davis/inprogress/QuakesEng3.html

188 https://sciencing.com/demagnetize-magnet-5071154.html

189 http://www.thunderbolts.info/webnews/121707electricsun.htm

190 https://www.sciencedaily.com/releases/2013/07/130718142726.htm

191 https://exoplanets.nasa.gov/news/110/a-frosty-landmark-for-planet-and-comet-formation/

192 https://www.space.com/17170-what-is-the-sun-made-of.html

193 https://physicsworld.com/a/radioactive-decay-accounts-for-half-of-earths-heat/

[194] http://aoi.com.au/bcw/FixedorExpandingearth.htm

[195] http://www.berkeley.edu/news/media/releases/2003/12/10_heat.shtml

[196] https://www.youtube.com/watch?v=JeMzOhoJpfw

[197] https://www.nasa.gov/audience/forstudents/5-8/features/space_gardens_feature.html

[198] https://van.physics.illinois.edu/qa/listing.php?id=1734

[199] https://www.enchantedlearning.com/geography/continents/Land.shtml

[200] https://www.terrapub.co.jp/journals/JO/pdf/4903/49030249.pdf

[201] https://www.ngdc.noaa.gov/mgg/ocean_age/data/2008/image/age_oceanic_lith.jpg

[202] https://web.archive.org/web/20100409010302/http:/www.expanding-Earth.org/page_19.htm

[203] http://www.naturalphilosophy.org/pdf/abstracts/abstracts_6518.pdf

[204] https://www.sciencedaily.com/releases/2008/04/080430112530.htm

[205] https://www.britannica.com/science/oceanic-crust

[206] https://oceanservice.noaa.gov/facts/oceandepth.html

[207] http://jrscience.wcp.muohio.edu/fieldcourses06/PapersMarineEcologyArticles/HydrothermalVent.html

[208] https://oceanservice.noaa.gov/facts/vents.html

[209] https://ngdc.noaa.gov/mgg/global/etopo1_ocean_volumes.html

[210] https://www.britannica.com/science/Jurassic-Period/Occurrence-and-distribution-of-Jurassic-rocks

[211] https://pubs.geoscienceworld.org/gsa/gsabulletin/article-abstract/62/9/1111/4461/geologic-history-of-sea-wateran-attempt-to-state

[212] https://www.nature.com/news/tiny-diamond-impurity-reveals-water-riches-of-deep-Earth-1.14862

[213] http://www.newscientist.com/article/dn25723-massive-ocean-discovered-towards-earths-core.html?utm_source=NSNS&utm_medium=SOC&utm_campaign=hoot&cmpid=SOC%7CNSNS%7C2013-GLOBAL-hoot

[214] http://www.huffingtonpost.com/2014/06/13/hidden-ocean-Earth-core-underground-video_n_5491692.html

[215] http://time.com/2868283/subterranean-ocean-reservoir-core-ringwoodite/

[216] http://www.livescience.com/1312-huge-ocean-discovered-Earth.html

[217] https://source.wustl.edu/2007/02/3d-seismic-model-of-vast-water-reservoir-revealed/

[218] https://sureshemre.wordpress.com/2014/05/04/huge-amount-of-water-in-earths-mantle/

[219] http://journal.frontiersin.org/article/10.3389/feart.2014.00038/full

[220] http://ngdc.noaa.gov/mgg/global/etopo1_ocean_volumes.html

[221] http://www.ngdc.noaa.gov/mgg/ocean_age/data/2008/ngdc-generated_images/whole_world/2008_age_of_oceans_plates.jpg

[222] https://web.archive.org/web/20130606230910/http:/www.ccmr.cornell.edu/education/ask/index.html?quid=888

[223] https://www.eas.cornell.edu/faculty-directory/bill-white

[224] https://www.velikovsky.info/in-the-beginning/

[225] http://www.ifiseeu.com/modern-mythology/Velikovsky-bio.htm

[226] http://the-eye.eu/public/concen.org/01052018_updates/True History of Atlantis %26 New Sumerian Artifacts %5Bpack%5D/ebooks/Worlds in Collision by Immanuel Velikovsky.pdf

[227] https://www.slideshare.net/operacrazy/revised-velikovsky-slideshow-54355793

[228] http://www.varchive.org/itb/ecocean.htm

[229] https://www.unm.edu/~toolson/salmon_osmoregulation.html

[230] https://www.sciencealert.com/Earth-s-water-didn-t-just-come-from-comets-says-new-study

[231] https://geology.com/stories/13/salt-domes/

[232] https://oceanservice.noaa.gov/facts/whysalty.html

[233] http://pubs.usgs.gov/dds/dds-067/CHB.pdf

[234] https://web.archive.org/web/20100124223253/http:/news.nationalgeographic.com/news/2006/04/0426_060426_dinos.html

[235] http://www.petroleum.co.uk/abiotic-oil-formation

[236] https://www.spacetelescope.org/news/heic1322/

[237] https://www.jpl.nasa.gov/news/news.php?release=2011-309

[238] https://solarsystem.nasa.gov/news/11197/methane-on-titan-and-enceladus-nature-vs-nurture/

[239] https://www.space.com/17470-neptune-moon-triton-subsurface-ocean.html

[240] https://uk.reuters.com/article/us-space-saturn-moon/saturn-moon-may-have-life-friendly-underground-ocean-scientists-idUKKCN0I52KJ20141016

[241] http://www.nasa.gov/press/2015/march/nasa-s-hubble-observations-suggest-underground-ocean-on-jupiters-largest-moon/

[242] https://sservi.nasa.gov/articles/nasas-hubble-observations-suggest-underground-ocean-on-jupiters-largest-moon/

[243] https://www.nature.com/news/2001/010726/full/news010726-12.html

[244] http://www.britannica.com/EBchecked/topic/176199/earthquake/247989/Shallow-intermediate-and-deep-foci

[245] http://www.csmonitor.com/Science/2014/0624/Alaska-earthquake-rings-the-Earth-like-a-bell

[246] http://www.rsc.org/Education/Teachers/Resources/jesei/fascinat/home.htm

[247] http://www.gns.cri.nz/Home/Learning/Science-Topics/earthquakes/Monitoring-earthquakes/Other-earthquake-questions/Does-the-Earth-really-ring-like-a-bell-after-a-big-earthquake

[248] http://www.nasa.gov/exploration/home/15mar_moonquakes_prt.htm

[249] http://www.lpi.usra.edu/lunar/missions/apollo/apollo_15/experiments/ps/

[250] http://www.auricmedia.net/wp-content/uploads/2015/01/Don_Wilson_Spaceship_Moon.pdf

[251] https://www.dinox.org/bookdetails.html

[252] http://www.sciforums.com/threads/newtons-shell-theorem-%E2%80%93-bad-mathematics-bad-physics.92918/

[253] http://galileo.phys.virginia.edu/classes/152.mf1i.spring02/GravField.htm

[254] http://www.math.ksu.edu/~dbski/writings/shell.pdf

[255] http://www.youtube.com/watch?v=c_H90oYLxqY

[256] https://amedleyofpotpourri.blogspot.com/2018/05/gausss-law-for-gravity.html

[257] https://science.nasa.gov/astrophysics/focus-areas/what-is-dark-energy

[258] http://hyperphysics.phy-astr.gsu.edu/hbase/orbv.html

[259] http://onlinelibrary.wiley.com/doi/10.1029/TC003i001p00045/abstract

[260] http://www.newscientist.com/article/dn24068-gravity-map-reveals-earths-extremes.html

[261] http://ddfe.curtin.edu.au/gravitymodels/GGMplus/hirt2013_ultrahighres_gravity.pdf

[262] http://onlinelibrary.wiley.com/doi/10.1002/cjg2.914/abstract

[263] https://www.amazon.com/s?k=Smart-Weigh-GEM20

[264] https://www.physicsforums.com/threads/dirac-and-gravity.36201/

[265] https://www.britannica.com/science/gravity-physics/Experimental-study-of-gravitation

[266] http://www.einstein-online.info/spotlights/scalar-tensor/

[267] https://www.youtube.com/watch?v=KnNUTOxHoto

[268] http://www.drjudywood.com/

[269] https://www.checktheevidence.com/wordpress/free-pdf-books/

[270] https://www.ancient-world-mysteries.com/360-days-Earth-year.html

271 https://www.youtube.com/watch?v=tBT8KyWVxj8

272 http://www.jpl.nasa.gov/missions/explorer-1/

273 http://www.thespacereview.com/article/1224/1

274 https://www.checktheevidence.com/wordpress/2010/04/26/wikipedia-censorship-of-9-11-evidence-and-related-legal-action-continues/

275 http://en.wikipedia.org/wiki/List_of_lunar_probes

276 http://science.nasa.gov/science-news/science-at-nasa/2006/28jul_crashlanding/

277 http://www.astronomy.ohio-state.edu/~pogge/Ast162/Intro/gravity.html

278 https://armaghplanet.com/whatever-happened-to-transient-lunar-phenomena.html

279 http://www.astrosurf.com/luxorion/ltp-r277-index.htm

280 http://science.nasa.gov/science-news/science-at-nasa/2014/13oct_moonvolcano/

281 http://photojournal.jpl.nasa.gov/catalog/PIA18821

282 http://www.space.com/27308-moon-ocean-of-storms-giant-rectangle.html

283 http://earthsky.org/space/new-ideas-about-origin-of-lunar-ocean-of-storms

284 http://www.space.com/26236-moon-far-side-mystery-maria.html

285 https://web.archive.org/web/20130815192131/http:/www2.astro.psu.edu/users/cpalma/astro10/class21.html

286 http://www.ast.cam.ac.uk/public/ask/solar-system

287 https://www.electricuniverse.info/

288 https://www.holoscience.com/wp/electric-galaxies/

289 http://www.ancientdestructions.com/wallace-thornhill/

290 http://www.holoscience.com/wp/sample-page/

291
https://www.amazon.co.uk/s?i=stripbooks&rh=p_27%3AWallace+Thornhill&s=relevancerank&text=Wallace+Thornhill&ref=dp_byline_sr_book_1

292 https://www.youtube.com/user/ThunderboltsProject

293 http://www.everythingselectric.com/birkeland-currents/

294 https://www.everythingselectric.com/eu-conferences/

295 https://www.holoscience.com/wp/electric-gravity-in-an-electric-universe/

296 https://www.youtube.com/watch?feature=player_embedded&v=34wtt2EUToo

297 https://www.pollacklab.org/jerry

298 https://www.nobelprize.org/prizes/medicine/1937/szent-gyorgyi/biographical/

299 https://www.gilbertling.org/

300 https://www.pollacklab.org/research

301 https://www.amazon.co.uk/Fourth-Phase-Water-Beyond-Liquid/dp/0962689548

302 https://www.metlink.org/wp-content/uploads/2013/06/PhysRev25_4_Nicoll.pdf

303 https://youtu.be/RnaG6T-C14Q?t=2194

304 https://checktheevidencecom.ipage.com/checktheevidence.com/audio/2016_10_09 - Fredrik Nygaard - Andrew Johnson - Dipole Gravity and Expanding Earth.mp3

305 http://www.universeofparticles.com/wp-content/uploads/2018/05/Universe-of-Particles.pdf

306 http://www.universeofparticles.com/

307 https://www.checktheevidence.com/wordpress/?s=nygaard&submit=Search

308 http://www.universeofparticles.com/hollow-planets/

309 https://www.welmec.org/fileadmin/user_files/pdf/gravity/gravity_info_Iceland.pdf

310 https://physicsabout.com/difference-coulombs-law-newtons-gravitational-law/

311 https://youtu.be/EckBfKPAGNM?t=2803

[312] http://espanol.apologeticspress.org/articles/140002

[313] https://www.britannica.com/science/redshift

[314] http://www.universeofparticles.com/mass-condensation/

[315] https://www.cs.unc.edu/~plaisted/ce/redshift.html

[316] https://www.checktheevidence.com/wordpress/2017/06/30/jupiters-large-and-diffuse-core-fredrik-nygaard/

[317] https://www.nasa.gov/mission_pages/juno/overview/index.html

[318] https://www.nature.com/news/jupiter-s-secrets-revealed-by-nasa-probe-1.22027

[319] https://www.youtube.com/watch?v=CvfFJiUWuDk

[320] https://www.checktheevidence.com/wordpress/2016/10/01/the-gravity-mystery/

[321] https://science.howstuffworks.com/antarcticas-spooky-cosmic-rays-might-shatter-physics-as-know-it.htm

[322] https://www.spiegel.de/international/spiegel/africa-s-new-ocean-a-continent-splits-apart-a-405947.html

[323] https://www.geolsoc.org.uk/Plate-Tectonics/Chap3-Plate-Margins/Divergent/Triple-Junction

[324] https://www.youtube.com/watch?v=3jpbArY2L78

[325] https://web.archive.org/web/20160318233013/http:/www.metaresearch.org/solar system/eph/eph2000.asp

Made in the USA
Monee, IL
07 July 2026

56550995R00100